# Climate Change and Risk Mitigation: Reducing Vulnerabilities and Enhancing Resilience

## CABI CLIMATE CHANGE SERIES

Climate change is a major environmental challenge to the world today, with significant threats to ecosystems, food security, water resources and economic stability overall. In order to understand and research ways to alleviate the effects of climate change, scientists need access to information that not only provides an overview of and background to the field, but also keeps them up to date with the latest research findings.

This series addresses many topics relating to climate change, including strategies to develop sustainable systems that minimize impact on climate and/or mitigate the effects of human activity on climate change. Coverage will encompass all areas of environmental and agricultural sciences. Aimed at researchers, upper level students and policy makers, titles in the series provide international coverage of topics related to climate change, including both a synthesis of facts and discussions of future research perspectives and possible solutions.

### Titles Available

1. Climate Change and Crop Production
   Edited by Matthew P. Reynolds
2. Crop Stress Management and Climate Change
   Edited by José L. Araus and Gustavo A. Slafer
3. Temperature Adaptation in a Changing Climate: Nature at Risk
   Edited by Kenneth Storey and Karen Tanino
4. Plant Genetic Resources and Climate Change
   Edited by Michael Jackson, Brian Ford-Lloyd and Martin Parry
5. Climate Change Impact and Adaptation in Agricultural Systems
   Edited by Jürg Fuhrer and Peter J. Gregory
6. Livestock Production and Climate Change
   Edited by Pradeep K. Malik, Raghavendra Bhatta, Junichi Takahashi, Richard A. Kohn and Cadaba S. Prasad
7. Climate Change and Insect Pests
   Edited by Christer Björkman and Pekka Niemelä
8. Climate Change and Agricultural Water Management in Developing Countries
   Edited by Chu Thai Hoanh, Vladimir Smakhtin and Robyn Johnston
9. Climate Change Challenges and Adaptations at Farm-level: Case Studies from Asia and Africa
   Edited by Naveen Prakash Singh, Cynthia. Bantilan, Kattarkandi Byjesh and Swamikannu Nedumaran
10. Climate Change Impacts on Urban Pests
    Edited by Partho Dhang
11. Climate Change Impacts and Sustainability: Ecosystems of Tanzania
    Edited by Pius.Z. Yanda, Claude G. Mung'ong'o and Edmund B. Mabhuye
12. Climate, Ticks and Disease
    Edited by Pat Nuttall
13. Climate Change, Adaptation and Gender: Policy, Practice and Methodological Underpinnings Mamta Mehar and Narayan Prasad
14. Climate Change and Sustainable Agro-ecology in Global Drylands
    Edited by Adel El-Beltagy, Rattan Lal and Kauser Malik
15. Climate Change and Food Security
    Edited by Sukanta Mondal, Pankaj Singh and Ram Lakhan Singh
16. Epigenetics for Climate-Smart and Sustainable Agriculture
    Edited by Jen-Tsung Chen
17. Climate Change and Risk Mitigation: Reducing Vulnerabilities and Enhancing Resilience
    Edited by Paul Barnes

# Climate Change and Risk Mitigation: Reducing Vulnerabilities and Enhancing Resilience

*Edited by*

**Paul Barnes**
*HEAL Senior Research, Policy and Practice Fellow, Centre for Environment and Population Health, School of Medicine and Dentistry, Griffith University, Australia.*

CABI

**CABI is a trading name of CAB International**

CABI
Nosworthy Way
Wallingford
Oxfordshire OX10 8DE
UK

Tel: +44 (0)1491 832111
E-mail: info@cabi.org
Website: www.cabi.org

CABI
200 Portland Street
Boston
MA 02114
USA

T: +1 (617)682-9015
E-mail: cabi-nao@cabi.org



A catalogue record for this book is available from the British Library, London, UK.

ISBN-13: 9781800624139 (hardback)
9781800624146 (ePDF)
9781800624153 (ePub)

DOI: 10.1079/9781800624153.0000

Commissioning Editor: David Hemming
Editorial Assistant: Emma McCann
Production Editor: Rosie Hayden

Typeset by Exeter Premedia Services Pvt Ltd, Chennai, India
Printed and bound in the UK by CPI Group (UK) Ltd, Croydon, CR0 4YY

# Contents

# Contributors


**Anshu Sharma:** Co-founder, SEEDS, New Delhi, India. anshu@seedsindia.org

**Bruna De Marchi**: SVT – University of Bergen, Norway. brunademarchi@gmail.com

**Christian Dimmer:** School of International Liberal Studies, Waseda University, Japan. christian.dimmer@waseda.jp

**Clifford Shearing:** Professor Emeritus, at Public Law, Faculty of Law University of Cape Town, South Africa and at the Centre for Criminology and Sociolegal Studies, University of Toronto, Canada; Honorary Professor, School of Regulation and Global Governance, The Australian National University, Canberra, Australia

**Enembe O. Okokon**: HEAL Global Research Centre, Health Research Institute, University of Canberra, Australia; National Centre for Epidemiology and Population Health, Australian National University, Australia; University of Calabar, Calabar, Nigeria. Enembe.Okokon@canberra.edu.au

**Esther Charlesworth:** School of Architecture and Urban Design, RMIT University, Melbourne, Australia. esther.charlesworth@rmit.edu.au

**Evelyne de Leeuw**: Canada Excellence Research Chair 'One Urban Health', University of Montreal, Canada. evelyne.de.leeuw@umontreal.ca

**Geoff Green:** Centre for Regional, Economic and Social Research, Sheffield Hallam University, UK. g.green@shu.ac.uk

**Haripriya Kesavan:** Spatial Planning and Product Technical Specialist, Resilience AI, New Delhi, India. haripriya@resilience360.ai

**John Fien:** School of Architecture and Urban Design, RMIT University, Melbourne, Australia. john.fien@rmit.edu.au

**Kurt Ackermann:** South African Urban Food and Farming Trust (SAUFFT), Cape Town, South Africa; Global Risk Governance Programme, University of Cape Town, South Africa. kurt@fairfood.org.za

**Mark Crosweller:** AFSM Founder and Director – Ethical Intelligence Pty Ltd, Australia. mark.crosweller@ethicalintelligence.com.au

**Michael Tong:** National Centre for Epidemiology and Population Health, Australian National University, Canberra, Australia. Michael.Tong@anu.edu.au

**Paul Barnes:** HEAL Senior Research, Policy and Practice Fellow, Centre for Environment and Population Health, School of Medicine and Dentistry, Griffith University, Australia. paul.barnes@griffith.edu.au

**Samhita R.:** CEO and Co-Founder, Innovation and Advocacy, Resilience AI, New Delhi, India. shradha@resilience360.ai

**Anshu Sharma:** Co-Founder, SEEDS, New Delhi, India. anshu@seedsindia.org

**Saul Roux:** Head of Environmental Strategy Implementation, City of Cape Town, South Africa. saulroux@gmail.com

**Scira Menoni:** DABC – Politecnico di Milano, Italy. scira.menoni@polimi.it

**Sotiris Vardoulakis:** HEAL Global Research Centre, Health Research Institute, University of Canberra, Australia. Sotiris.Vardoulakis@canberra.edu.au

**Tamsin Faragher:** City of Cape Town, Risk and Resilience Department, Cape Town, South Africa; African Centre for Cities, University of Cape Town, South Africa

# Foreword

Paul Barnes has brought together an impressive group of experts to address one of the most urgent challenges of our time: navigating the cascading risks of climate change and natural disasters. Particularly noteworthy is the substance and the breadth of perspectives that are presented in this important volume.

Dr Barnes has long advocated for governments to comprehensively address disaster risk, including systemic risks. He has also emphasized the importance of breaking down institutional silos and the need for interdisciplinary approaches to better understand and respond to risks. The chapter authors, drawing upon scientific, technical and policy research, offer a range of innovative viewpoints touching on these themes and others, including the need for new governance approaches that blend science, policy and community insight.

From multi-sectoral approaches to the ethical dimensions of climate adaptation, to discussions on urban resilience and health impacts, the contributions in this book together enhance our understanding of complex risks and how to address them more effectively.

One of the strengths of this volume is that it recognizes that disaster risk is rarely confined to one domain. Climate-fuelled hazards can spill rapidly across boundaries, triggering cascading disruptions. Australia's Black Summer bushfires in 2019 for example vividly demonstrated that these systemic impacts are now appearing on a national scale. The fires destroyed over 20 million hectares, with major cascading effects on infrastructure, health, biodiversity, tourism and the economy. By bringing these interconnections into sharper focus, the authors remind us that responding effectively requires not just technical siloed fixes, but also the integration of social, ethical and political considerations within risk governance.

This book also makes clear that solutions cannot be imposed from the top down. Effective adaptation and resilience require approaches that engage communities directly, drawing on local knowledge and experience, while also ensuring that decision makers have access to the best available science. This balance between expert analysis and community participation is often difficult to achieve but, particularly with climate change, it is becoming increasingly essential.

Finally, the diversity of perspectives gathered here illustrates the value of cross-disciplinary collaboration. By sharing insights from engineering, public health, ethics, urban planning and disaster management, this collection demonstrates the importance of working across disciplinary, jurisdictional and institutional silos. It not only helps us to diagnose the challenges we face more clearly, but also shows how collaborative, innovative thinking can contribute to building more resilient societies.

In short, this book offers a timely resource for those seeking practical and inclusive approaches to resilience in a rapidly warming world.

**Dr Robert Glasser**
Honorary Professor, Institute for Climate, Energy & Disaster Solutions, Australian National University and former UN Secretary General's Special Representative for Disaster Risk Reduction.

# Introduction

**Paul Barnes***
*HEAL Senior Research, Policy and Practice Fellow, Centre for Environment and Population Health, School of Medicine and Dentistry, Griffith University, Australia*

As the world grapples with an evolving disaster risk landscape, it has become increasingly evident that the impacts of climate change are not only far-reaching but also manifest in complex ways. The cascading effects of natural hazards, exacerbated by global warming effects and destructive weather events, pose unprecedented challenges to governments, businesses and communities alike. The veracity of this evolving landscape is well evidenced in recent global research efforts (IPCC, 2022, 2023).

The extent of resulting disturbances is often difficult to anticipate, let alone control, regularly extending beyond simple disruptions to encompass series of systemic failures and related impacts – often appearing concurrently as emergent *network effects* – within and across essential services, affecting life, property and the environment on both regional and global scales. Such events have been categorised as 'outside of the box', 'too fast' and 'too strange' (Lagadec, 2004).

The complexity and uncertainty surrounding these realities make the work of policy governance and agencies tasked with supporting adaptation outcomes, risk mitigation, urban planning and community resilience even more challenging and require innovation in thinking, action and governance (Topper and Lagadec, 2013).

In 2022, Selwin Hart, Special Adviser to the UN Secretary-General on Climate Action and Assistant Secretary-General for Climate Action, noted that the impacts of climate change will be unevenly distributed, with those living in diverse landscapes such as Central and South America, Central, East or West Africa, South Asia and small island developing states being up to 15 times more likely to experience severe consequences due to cascading climate effects (UNDRR, 2022).

Despite the significant exposure to harm faced by these regions, Hart pointed out that only about 20–25% of all climate-related investments were directed towards adapting to these changes or enhancing the readiness of communities and public and private sector institutions or strengthening the resilience of essential services. While there are indications that this imbalance is being addressed, historically questions have been raised about the balance between mitigation and adaptation effectiveness and adequacy of funded solutions, as well as the quality of decisions surrounding their selection and implementation strategies (UNDRR, 2021).

In a recent report, the World Meteorological Organization stated that 2023 was the warmest year on record to date, a trend continued into 2024. Climate extremes have become a regular lived experience for humans over increasingly longer durations and at greater intensities (WMO, 2024).

*Corresponding author: paul.barnes@griffith.edu.au

 *Climate Change and Risk Mitigation: Reducing Vulnerabilities and Enhancing Resilience* (ed. P. Barnes)
DOI: 10.1079/9781800624153.intr

Across the period from 2020 to mid-2024, floods remained the most frequently reported disaster. However, heat-related hazards became the leading cause of deaths, accounting for 57% of the total of reported weather-, water- and climate-related deaths globally. Storms resulted in the greatest economic losses, contributing to 59% of the total recorded cost (WMO, 2024). The effect of recent intense rainfall and subsequent flooding across Europe (late 2024) may indicate a continuation of this trend.

Historically, from relatively benign times, the responsibility for regulatory decisions about societal safety has rested in the hands of professionalised bureaucracies, scientific specialists and elected officials (Plough and Krimsky, 1987). While lay individuals can take actions such as evacuating their homes in the face of an approaching cyclone or choosing to participate in certain dangerous or hazardous pursuits, they cannot avoid the consequences of regulatory decisions made by others.

The role of government in preventing, or at least mitigating the severity of the impacts of natural hazards, is critical in that varieties of the 'modern state' as functional providers of public administration and governance derive validity from the promise that it is an effective (and preferred) form for the provision of safety (and certainty) in society. This role is made more important because, in representative democracies, it is a reality that many decisions with potential to affect our lives are made by others.

Such power differentials can manifest as intractable concerns among members of the public who have found themselves distant from the decision-making processes (Luhmann, 1990). It is for this reason that government legitimacy is precisely so important, as it highlights the ways authorities draw upon their sphere of influence to make representative decisions and reify state, society, mutual trust and support.

This pattern of governance is well evidenced in the evolution of health and safety knowledge and related standards and in the professionalisation of risk and safety specialists operating from within institutions (Heimer, 1988, p. 512). A key contention in this context is the benefit of decisions on societal safety made at a centralised, institutional level rather than at community or individual levels (MacLean, 1982; Clarke, 1988).

This situation raises the question of where the responsibility for decisions about preparedness and risk mitigation in the face of climate change should lie. The complexities and uncertainties associated with adapting to and mitigating the impacts of a changing climate may require a more inclusive and participatory approach to decision making.

Commentary about the evolution of professionalised risk assessment has been an ongoing point of discussion over many years, particularly the notion of the inability of a (generalised) public to understand technical knowledge and nuanced evidence (Bradbury, 1989). For such a technical elite, the inability of a public audience to understand (accept) institutional risk estimates may be seen as analogous to the situation faced by engineering students in their first semester: in a sotto voce form, 'they are ignorant but well-intentioned, hardworking but without a clue' (Beck, 1992, p. 58).

While these views about knowledge and advocacy gaps can be seen to have emerged from the professionalisation of public health, engineering science and safety navigating the evolving disaster-risk landscape in an era of cascading crises may require a fundamental shift away from professionalised risk governance. It would seem to demand a more inclusive, adaptive and integrated approach that engages communities, stakeholders and citizens in the framing and management of the potential for harm while also drawing on the best available science, policy analysis and ethical considerations to inform decision-making processes.

A more diffuse approach to understanding threat environments from the perspective of those at risk of harm from existential threats, naturally sourced disasters and climate change-related issues needs to consider agility and wider involvement in decisions about adapting to the global uncertainties that are in our present and future. Such contexts might in fact be messy, lacking technical efficiency.

This volume explores – approaches to and dialogue on – the framing of risk mitigation options from a range of different yet complementary perspectives. These range from considering increasing community agency in cases of extreme climate uncertainty, improving decisions with detailed policy analysis and ethical consideration, appraising selected knowledge sets underpinning compounding weather events, options for building (re)design, design and functional adaptations in human settlements and threats to human health under conditions of excess heat and humidity.

Chapter 1, Breaking Silos: Multi-Sectoral Approaches to Mitigating the Cascading Impacts of Climate and Natural Hazards, by Paul Barnes, discusses a view that we cannot solve our (current) problems with the same thinking used in their creation. By analogy this suggests that for meta-scale issues such as climate change and weather-driven catastrophes, breaking patterned (siloed) thinking is a viable way to proceed. This chapter considers the importance of accessing a 'requisite variety' of concepts and framing to effectively scope options that can reduce the impacts and cascading effects of a changing and less predictable climate. It will also suggest that pursuing multidisciplinarity, aimed at delivering transdisciplinary outcomes, is a goal worth pursuing.

Chapter 2, Who Decides? Choosing Adaptation Tools for Climate Risk. A Post-Normal Science Problem, by Bruna De Marchi and Scira Menoni, advocates for the application of the Post-Normal Science (PNS) approach in the study and management of hazards and disasters, particularly in the context of climate change adaptation. PNS addresses policy-relevant problems characterised by uncertain facts, disputed values, high stakes and urgent decisions, and suggests that the decision process must include inputs from various disciplines and 'lay actors' in an 'extended peer community'. The authors highlight the importance of knowledge co-production, and the inclusion of different forms of knowledge (scientific, legislative, organisational and common) in framing complex problems related to disaster risk reduction and climate change adaptation. They argue that the PNS approach offers a problem-solving strategy that emphasises the quality of the decision process, grounded in criteria of inclusiveness, equity and fairness, rather than the search for a unique and definitive truth.

Chapter 3, Ethics in Climate Risk Mitigation and Policy Development, by Mark Crosweller, explores the ethical challenges faced by disaster management leaders when advising governments on policies and strategies related to climate change-driven disasters. The author highlights the increasing frequency and intensity of compounding and cascading disruptions driven by climate change, which have resulted in continuous or contiguous disaster events with large-scale impacts on lives and livelihoods. Disaster management frameworks and reviews often fail to fully consider the social, economic and structural factors that contribute to exposure and vulnerability, leading to a highly contested concept of shared responsibility between governments, businesses, non-governmental organisations, communities and individuals.

Chapter 4, Addressing the Wicked Problem of Systemic Disaster Risk in the Anthropocene, by John Fien and Esther Charlesworth, begins with examining how environmentally sourced disasters represent a microcosm of the broader 'polycrisis' facing humanity. They argue that living within the 'Anthropocene' – a geological epoch where human activity influences planetary changes – creates complex, interlocking risk exposures that are increasingly unpredictable and severe. They contend that traditional approaches to disaster management have focused on engineering solutions like sea walls and levee banks, which may mitigate immediate impacts but fail to address the root socio-economic and political drivers of vulnerability. The authors outline three broad principles for addressing disaster risk: developing a new social contract based on shared values; creating agile, adaptive governance arrangements; and guiding decisions through evidence-based knowledge, enabled by the connection of theoretical frameworks (the Anthropocene, complex systems, wicked problems) with practical approaches (systemic design, participatory strategies). The chapter concludes by proposing 'systemic design' – an integration of systems thinking and design thinking – as a framework for addressing disaster risk management in the Anthropocene. Such an approach involves seven iterative steps: framing the system, listening to the system, understanding interconnections, defining desired futures, exploring possibilities, designing intervention strategies and implementing with continuous monitoring. By treating interventions as experiments to be closely observed and revised, systemic design can enable communities to co-evolve with natural systems rather than attempting to control them.

Chapter 5, Is Better Policy Enough? Governance of Urban Energy Transitions in the Global South, by Saul Roux and Clifford Shearing, explores the conditions, pathways and governance of energy reconfigurations in Cape Town intersecting challenges related to climate change, poverty and urbanisation – and the multiple crises of the Anthropocene. It focuses on the role of value-sets and networks in driving socio-technical regime change. The analysis is situated within the context of debates related to the conditions that make reconfigurations possible, contestation within socio-technical regimes, nested multi-level governance, the politics of transitions and the role of collective actors or networks and state actors in transitions.

Chapter 6, Hotter and Damper: Dealing with the Environmental Health Impacts of Changing Weather and Climate, by Enembe Okokon, Michael Tong and Sotiris Vardoulakis, examines the effects of global warming on human thermal comfort and health. The authors explain that rising temperatures increase the moisture-carrying capacity of the atmosphere, leading to more intense precipitation and floods. The combination of rising temperatures and humidity negatively impacts human thermal comfort and increases the incidence of heat strain around the world, particularly in urban areas due to the urban heat island effect. The authors propose a multi-pronged, adaptive response to increasing environmental temperatures and humidity, involving diverse stakeholders and sectors. Potential solutions include reimagining urban morphology, prioritising passive cooling building designs, providing heat warning systems and health education, and establishing urban cooling centres. The text also emphasises the importance of monitoring and evaluation to assess the effectiveness of adaptation strategies and inform future planning. The chapter underscores the need for strong collaboration across sectors and policy portfolios, coupled with meaningful community engagement, to strengthen resilience to a hotter and wetter climate.

Chapter 7, Healthy, Resilient Cities, by Evelyne de Leeuw and Geoff Green, advocates for a holistic approach to building urban health resilience in the face of global challenges, particularly climate change. The authors argue that focusing solely on health may be counterproductive due to the dominance of the medical–industrial complex and the practice of 'healthism'. Instead, they propose considering the complex interactions between the geosphere, hydrosphere, atmosphere, biosphere and anthroposphere to promote a more comprehensive understanding of urban resilience. To build healthy and resilient cities, the authors advocate for a transdisciplinary approach that breaks free from disciplinary atomisation, hyperspecialisation towards technocratic approaches and expert-driven solutions. They argue for embracing a more integrated and engaged approach that recognises the interconnectedness of all spheres of life on the planet. The Age-Friendly Environments in Europe (AFEE) project is presented as an example of a framework that can inform urban resilience efforts in the face of climate change, demonstrating the potential for a more integrated and engaged approach to urban resilience for health and wellbeing in the face of polycrises. It concludes by emphasising a need for a context-specific, transdisciplinary and integrated approach to building healthy and resilient cities. The authors argue that effective urban resilience requires a shift away from a health-centric perspective and towards a more comprehensive understanding of the complex interactions between humans, the environment and governance structures. By embracing this holistic approach, cities can better prepare for and adapt to the challenges posed by global-scale change.

Chapter 8, by Ackermann and Faragher, present a case study of the Langa Agrihub, in Cape Town, South Africa, which was established to support a network of community food gardens and test theories about the role of Agrihubs in building community resilience. From a wider context, Cape Town faces various stresses, including income inequality, poverty, food insecurity, violent crime, joblessness and the impacts of climate change, which amplify vulnerability in under-resourced communities. Agrihubs have been promoted to address socio-economic challenges and contribute to a different approach to food production and nutrition that may guard against the emergent impacts of climate change on food security. This case study illustrates the potential of functioning Agrihubs to contribute to economic wellbeing, physical and mental health, safety, social wellbeing, civic engagement, urban green spaces, biodiversity, environmental quality and community resilience. The case study highlights the importance of community partnerships, farmer involvement and a holistic approach to addressing challenges and achieving positive outcomes across multiple dimensions of resilience.

Chapter 9, Adaptive Urban Resilience Using Artificial Intelligence, by Haripriya Kesavan, Samhita R and Anshu Sharma, examines how India's cities are facing increasingly severe and frequent heatwaves that are having disproportionate impacts on vulnerable communities living in informal settlements and densely built urban areas. Factors like the urban heat island effect, lack of green spaces, poor ventilation and heat-absorbing building materials are intensifying the dangers of extreme heat, especially for low-income populations with limited access to cooling and health care. The authors detail the use of AI-based risk assessment tools to enable targeted, localised heat mitigation efforts, building true climate resilience by collaboration across stakeholders, integration of technology with existing systems and knowledge, and a strong commitment to community engagement and empowerment. A 'human-in-the-loop' approach leveraging AI's analytical power in concert with decisive human action is crucial for Indian cities to effectively tackle the complex challenges posed by a warming climate. The authors highlight the potential of AI-based tools to enable hyperlocal, building-level heat risk assessments that can inform targeted interventions and resource allocation to mitigate the impacts of heatwaves. For example, the AI showcased here - the ResSolv tool - uses geospatial data and building characteristics to map heat risk at a granular level. Analysing a building's embodied and operational energy can help identify opportunities for passive cooling strategies to reduce indoor heat stress.

Chapter 10, Tokyo's Perpetual Resilience Project: Between Local Knowledges and Universal Modernist Concepts, by Christian Dimmer, provides a critical examination of Tokyo's urban resilience strategies, particularly focusing on the recently launched Tokyo Resilience Project (TRP). The author traces the evolution of Tokyo's approach to disaster resilience from the pre-modern era, characterised by a distinct 'disaster culture' that accepted disasters as inevitable and focused on swift response and recovery, to the modern era, which saw a shift towards reliance on expert knowledge and engineered solutions for disaster mitigation, leading to a disengagement of non-expert citizens. The author argues that the TRP displays a static, modernist understanding of the city, contrasting with the ecological, interdependent and adaptive view of cities proposed by Jane Jacobs and others. The Japanese term *kyōjin* used in the TRP suggests a focus on engineered resilience rather than ecological resilience, which emphasises adaptation to change. The TRP's expert-driven approach may be efficient but lacks the democratic governance and citizen participation necessary for deep-rooted disaster preparedness. Dimmer highlights the need for Tokyo to shift from a static, engineering-focused approach to urban resilience towards a more dynamic, participatory and ecologically informed strategy that empowers citizens and fosters community resilience in the face of growing climate change-related challenges.

Throughout the book, the authors draw on a wide range of disciplinary perspectives and methodological approaches, including post-normal science, policy analysis, ethical consideration and context-specific case studies. By bringing together these diverse perspectives and methodologies, the book aims to promote a more nuanced and comprehensive understanding of the complex challenges and opportunities for effective risk governance in an era of cascading crises.

Ultimately, the book argues that navigating the evolving disaster-risk landscape requires a fundamental shift in the way we think about and approach risk governance. It calls for a more inclusive, adaptive and integrated approach that engages communities, stakeholders and citizens in the framing and management of risks, while also drawing on the best available science, policy analysis and ethical considerations to inform decision-making processes.

In contributing to the ongoing dialogue on effective risk governance in an uncertain future, this book aims to advance our understanding of the challenges and opportunities for building more resilient, equitable and sustainable communities in the face of an uncertain and rapidly changing world.

## References


Beck, U. (1992) *Risk Society: Towards a New Modernity*. Sage Publications, London.

Bradbury, J.A. (1989) The policy implications of differing concepts of risk. *Science, Technology, & Human Values* 14(4), 380–399.

Clarke, L. (1988) Explaining choices among technological risks. *Social Problems* 35, 22–35.

Heimer, C.A. (1988) Social structure, psychology, and the estimation of risk. *The Annual Review of Sociology* 14, 491–517.

IPCC (2022) *Climate Change 2022: Impacts, Adaptation and Vulnerability – Summary for Policymakers*. IPCC, Geneva, Switzerland. Available at: https://www.preventionweb.net/media/78315/download?startDownload=20241116 (27 February, 2025).

IPCC (2023) *Climate Change 2023: Synthesis Report*. IPCC, Geneva, Switzerland. doi:10.59327/IPCC/AR6-9789291691647

Lagadec, P. (2004) Crisis: A watershed from local, specific turbulences, to global, inconceivable crises in unstable and torn environments. Future Crises, Future Agendas (An Assessment of International Crisis Research), International Workshop, November 24–26, 2004, Sophia-Antipolis, Nice, France. Available at: https://www.patricklagadec.net/fr/pdf/Sophia_24-11-04.pdf (27 February, 2025).

Luhmann, N. (1990) Technology, environment and social risk: A systems perspective. *Industrial Crisis Quarterly* 4(3), 223–231.

MacLean, D. (1982) Risk and consent: Philosophical issues for centralized decisions. *Risk Analysis* 2(2), 59–67.

Plough, A. and Krimsky, S. (1987) The emergence of risk communication studies: Social and political context. *Science Technology and Human Values* 12(3–4), 4–10.

Topper, B. and Lagadec, P. (2013) Fractal crises – a new path for crisis theory and management. *Journal of Contingencies and Crisis Management* 21(1), 4–16.

UNDRR (2021) Policy brief – disaster risk reduction and climate change. Available at: https://globalplatform.undrr.org/sites/default/files/2022-02/UNDRR%20%282021%29%20Policy%20brief%20-%20Disaster%20risk%20reduction%20and%20climate%20change.pdf?startDownload=true (27 February, 2025).

UNDRR (2022) Urgency, Ambition, Political Commitment, and Follow-Through: A High-Level Dialogue on Strengthening Disaster and Climate Risk Governance. Available at: https://www.undrr.org/news/urgency-ambition-political-commitment-and-follow-through-high-level-dialogue-strengthening (27 February, 2025).

WMO (2024) *2024 State of Climate Services: Five-Year Progress Report (2019–2024)*. WMO-No. 1363. World Meteorological Organization, Geneva, Switzerland. Available at: https://library.wmo.int/viewer/69061/?offset=#page=1&viewer=picture&o=bookmark&n=0&q= (accessed July 10, 2025).

# 1 Breaking Silos: Multi-Sectoral Approaches to Mitigating the Cascading Impacts of Climate and Natural Hazards

**Paul Barnes***

*HEAL Senior Research, Policy and Practice Fellow, Centre for Environment and Population Health, School of Medicine and Dentistry, Griffith University, Australia*

**Abstract**

This chapter examines the inadequacy of traditional siloed approaches to addressing the complex, cascading impacts of climate change and natural hazards. It argues that the interconnected nature of contemporary climate-driven disruptions requires a fundamental paradigm shift toward multi-sectoral, collaborative problem-solving strategies. Using the concept of 'polycrises' – where multiple, interconnected crises interact to produce cumulative effects greater than their individual impacts, it draws on cybernetics theory, particularly Ashby's Law of Requisite Variety, to argue that institutional responses to preparing for disruptions must match the complexity and diversity of climate challenges through requisite variety in institutional expertise and approaches applied. A central framework presented is the 'crisis continuum', used to map the progression from preconditions and triggering events through emergent disturbances to primary impacts and ongoing disruptions. This framework facilitates understanding of how complex crises triggered by climate-related disruptions can evolve with impacts propagating across time scales and through vulnerable systems. The chapter advocates for the use of transdisciplinary teams integrating experts from social sciences, natural sciences, technology, policy, and communication fields. The effectiveness of such teams requires sophisticated integration mechanisms, including shared conceptual frameworks, structured decision making processes, and operational protocols that accommodate multiple disciplinary perspectives. Rather than reactive responses, the chapter emphasises developing anticipatory approaches to risk governance facilitated by this team environment. A final goal is the creation of institutional capabilities that can proactively address the emergent properties and interdependencies characterising modern socio-technical-environmental systems. This represents a shift from managing individual hazards to the collective governing of systemic threats within an increasingly complex global environment.

## Introduction

Humans are ingenious. They have a propensity, proven through time, to be problem solvers. This trait is even more useful if we consider that humans, and by extension society, contribute to conditions from which many of its pressing problems emerge. Albert Einstein is often credited with saying, 'We cannot solve our problems with the same thinking we used when we created them' (Calaprice, 2011). This quote, although not definitively attributed to Einstein, addresses a central factor in complex climate change and weather-driven catastrophes both in our recent past and into a collective future.

*Corresponding author: paul.barnes@griffith.edu.au


DOI: 10.1079/9781800624153.0001

The impacts of climate change and natural hazards are far-reaching and interconnected, affecting various sectors of society, including health, agriculture, infrastructure and the economy. Mora *et al.* (2018) provides traceable evidence for 467 pathways by which human health, water, food, economy, infrastructure and security have been and will be impacted by climate changes such as greater warming and humidity, extended heatwaves and increased precipitation, drought, floods, fires and storms, as well as sea-level rise and changes in natural land cover and ocean chemistry. These connected impacts highlight the need for a multi-sectoral, collaborative approach to mitigate risk and build resilience (UNDRR, 2019)

Such cascading impacts are prime examples of meta-scale problems that demand a paradigm shift in our approach to problem-solving (IPCC, 2022) and, ideally, draw on the knowledge and insights from fields such as climatology, ecology, engineering, economics and social sciences. In doing so we may gain a more holistic understanding of the issues at hand and identify potential synergies and trade-offs needed for effective risk governance (Bammer, 2013).

Climate-driven disturbances are systemic issues requiring that as a collective we break free from the confines of traditional, siloed thinking. Reyers *et al.* (2018) suggest that realising the importance of understanding the cross-scale dynamics of such catastrophic impacts (local to global) is also critical, particularly with impacted systems prone to non-linear changes and shifts in function. They further suggest that transformation (of thinking) rather than just adaptation is needed.

Options for mitigating harm from such an impact landscape derive from different areas of science, policy and legislative practice. Addressing historically embedded barriers, silos of excellence, that impede collaborative thought, effective communication and coordination between disciplines is required before harnessing the power of diverse perspectives and expertise needed to support development-viable solutions (Stember, 1991).

While pointing to 'multidisciplined' solutions might seem glib, actual steps to delivering silo-busting options to mitigate and adapt (where deemed possible) to both expected and emergent impacts of warming, aberrant weather and related natural hazards can remain hard to achieve. This chapter examines opportunities to limit siloed thinking in support of anticipating and mitigating the disturbances from climate change and weather-driven catastrophes.

Given the scale and distribution of disturbances triggered by climate change, this chapter begins with consideration of 'impact landscapes' in a context of polycrises – offering an approach to mapping types of expertise, using a crisis continuum framework (incorporating the notion of tipping points), to expand discussion about options to mitigate the impact of climate disturbances. It suggests there is benefit from seeking to bring together a *variety* of expertise needed to create options for novel approaches to thinking and effective and agile practices and policy development.

## Polycrisis: Fad or Useful Construct

The concept of the polycrisis has emerged in both popular and academic literature as a framework for understanding the complex, interconnected disruptions. Lawrence defines a 'global' polycrisis as 'the causal entanglement of crisis in multiple global systems in ways that significantly degrade humanity's prospects' (Lawrence, 2024a, p. 1). So a combined effect of entangled crises has cumulative impacts that can be different from and generally greater than those crises if manifested individually in isolation from one another. And because it may have no single underlying cause, the polycrisis operates as an emergent phenomenon that 'requires' strategies and frameworks that are multidisciplinary, systemic and cross-sectoral, as efforts to reduce one problem can easily worsen others.

Commentary from the United Nations Office for Disaster Risk Reduction (UNDRR) in the Global Assessment Report (GAR) on Disaster Risk Reduction suggests that 'surprise' is the new normal in an evolving global threat landscape (UNDRR, 2019). This new form derives from non-linear change manifesting across socio-technical, environmental and economic systems and over multiple dimensions and scales, more quickly and surprisingly than may have previously been thought possible.

The authors of the GAR promote the relevance of the metaphor of civilisation becoming a 'super-organism' through global interconnectedness of highly interdependent social, technical and biological systems, which is both philosophically attractive and technically challenging from a risk assessment and mitigation perspective. While such ideas are not new (see, for example, Teilhard de Chardin and Vladimir Ivanovich Vernadsky's coverage of the noosphere or analytical frames from 'Deep Ecology' among many), their application to field of disaster risk and impact reduction arguably is.

Mark *et al.* (2024) define a polycrisis as a space in which multiple, macroregional, ecologically embedded and deeply interconnected institutions and systems face high uncertainty across socio-economic, political and other dimensions. Mark *et al.* (2024) further suggest a range of challenges:

- The irreversible nature of many current environmental changes means that 'unlike in previous eras, many of the biophysical stressors currently faced are simply irreversible' (p. 15).
- The increasing interconnectedness between social systems creates feedback loops where 'the outcome from one crisis or shock impacts and bolsters others, leaving limited time for recovery before the next one hits' (p. 15).
- The erosion of social trust and institutional capacity makes coordinated responses more difficult, as 'trust and confidence in government and science limits taking collective action across society' (p. 22).

The authors identify several key characteristics that distinguish polycrises from other types of crises. First, they emphasise that modern decision makers face fundamentally different challenges than those faced by their predecessors because today's crises are deeply interconnected and operate across multiple systems simultaneously. As they note, 'Decision-makers in the 2020s no longer face individual shocks and crises that can be handled in isolation with a single-systems focus' (p. 2). They suggest recent crises present unprecedented challenges that require new frameworks of understanding and response. Further, as Tooze (2022) states, 'the shocks are disparate, but they interact so that the whole [effect] is even more overwhelming than the sum of the parts.'

Some pundits have questioned the authenticity of the term polycrisis. Andreas Kluth in the *Washington Post* (Kluth, 2023) wondered about the 'value in creating apocalyptic neologisms' that might not assist dealing with the issues that underpinned the term. Similarly, Drezner (2023) referred to it as the 'buzzword of the moment'. Kluth (2023) goes further to ask if polycrisis blends into the 'permacrisis' (Collins Dictionary Word of the Year in 2022) or might overtake the 'mega-threats' out there or might in retrospect be seen as the familiar 'perfect storm'.

Of course, while journalistic endeavours attempting to bring degrees of sarcasm and humour provide amusement, they do little to assist with clarity for consumers about the issues at hand or on choices for self-protection or understanding about options for attenuating the cascading and cumulative impacts noted. Interestingly, Mocatta *et al.* (2024) report findings from surveys of 744 journalists from 102 countries indicating that up to 68% reported they never or rarely used the term 'global polycrisis' in their reporting. Respondents also expressed belief that while they do not believe their audiences understand the concept, there was reader interest in it.

Enhancing the sophistication of explanations can only go so far. Mark *et al.* (2024) suggests a deeper analysis by emphasising a need to differentiate 'polycrises' from related concepts such as: cascading disasters (e.g. Japan's 2011 triple disaster); compound disasters (e.g. the overlap of COVID-19 and locust swarms in Africa) and recurring acute disasters (e.g. repeated seasonal disruptions from wildland fires and hydrometeorological events).

Lawrence *et al.* (2024b) also point to the importance of recognising recent conceptual thinking which emphasises 'the causal entanglement of crises in multiple global systems in ways that significantly degrade humanity's prospects' and maintain that 'when crises interact, they reshape and intensify one another to produce harms both greater than and different than the sum of the harms they would produce separately.'

Notwithstanding degrees of scepticism in the media about exciting new names for things, a reality we all face is the obvious high frequency of a myriad of instances of widespread failure of and disturbances to human, environmental and technical systems over recent times: some driven by climate and weather variability and others potentially emerging from the complexity of socio-technical systems themselves – as regular phenomena. This sense of *regularity* is not in reference to how frequent surprises might be but suggestive, as defined by Perrow (1999), of the complex interactivity intrinsic to the systems themselves that leads to conditions where component failure is more likely than not.

## Deconstructing the Emergence of Crises

With the impact landscape of polycrises being global and reactive to seasonal, temporal and cross-scale effects, it is not unreasonable to marvel at the complexity and enormity of deciding how to even begin addressing the challenges of mitigating its impacts and the wider risk governance options might be considered. There are deeper dialogues that might be held about the interplay between natural and technical threats, but a specific focus here is how climate and weather-related disturbances (within a polycrisis landscape) emerge or are triggered.

Raymond *et al.* (2020) refer to the idea of 'connected extreme events' in understanding and managing the complex, interrelated impacts of climate extremes. They define such connected extremes as combinations of events whose impacts are amplified by physical and societal linkages, distinguishing them from compound events based solely on physical drivers. Moreover, the interconnectedness of climate-related hazards necessitates a systems-focused perspective that recognises how cascading impacts can emerge from seemingly isolated events. For instance, flooding can lead to infrastructure damage, which in turn disrupts supply chains and affects public health (Zebisch *et al.*, 2022).

One major focus of research in the past decade has been the identification and characterisation of potential 'tipping points' in the Earth's climate system. Researchers have over a considerable time explored the interconnection between different elements and the potential for cascading effects. Extensions of this work have also introduced the concept of a 'tipping cascade', where the emergence of one tipping point could trigger a sequence of other tipping points, leading to a rapid and irreversible shift in the Earth's climate system. Lenton *et al.* (2019) investigated the potential for such interactions between tipping elements, highlighting the need for a more integrated, systems-level approach to understanding likely impacts from climate change (see also Werners *et al.*, 2013; Steffen *et al.*, 2018; Lenton *et al.*, 2019).

Lenton *et al.* (2008) provided a seminal review of tipping elements in climate systems, highlighting a range of potential transition foci, including loss of arctic sea ice, the Greenland ice sheet and the Amazon rainforest. Subsequent studies have further refined the understanding of these tipping elements, their thresholds and their potential impacts (Huntington *et al.*, 2012; Levermann *et al.*, 2012; Schellnhuber *et al.*, 2016; Lenton *et al.*, 2019).

Accurate assessments of such thresholds and timescales associated with tipping points, however, remain elusive. Schellnhuber *et al.* (2016) have emphasised the need for improved monitoring and early warning systems to detect the nearness of transitions. Additionally, Kriegler *et al.* (2009) highlighted the challenges in defining and identifying tipping points, given the complexities and uncertainties inherent in the climate system(s) where minor changes in key variables might trigger emergent, potentially irreversible impacts that constitute new norms.

Timing between a tipping point and any subsequent change may be gradual or more rapid. Such a shift of a system between alternate regimes intellectually has links to catastrophe theory. The complex nature of these interconnections and the potential for surprises and non-linear responses pose significant challenges for predicting and managing the associated disruption(s) (Reyers *et al.*, 2018). Kopp *et al.* (2016) emphasised the need for improved models and scenarios that capture the potential for cascading effects and abrupt changes. Recent attention to biophysical thresholds and tipping points, described above, suggests that it may be important to explore the existence of such phenomena in the social and political systems within which actual adaptation decisions play out.

Understand the dynamics of climate thresholds, tipping points and impacts, in detail, is beyond the purpose of this work. What is pertinent here is consideration of what institutions might do to apply to these issues the knowledge gained over the many years.

Milkoreit *et al.* (2018) provide detail on the evaluation of the concept of tipping points, highlighting that the concept has deep historical roots in mathematics and chemistry, first appearing in scientific papers by Hoadley (1884) and Poincaré (1885), where it was described mathematically as a bifurcation – a qualitative change in a system. Bifurcation theory continues to be used today in mathematics, physics and complex systems science and in examination of issues linked to complex adaptive systems and applications in catastrophe theory.

Seminal work referenced by Milkoreit *et al.* (2018) details the later use of the term in political science through Grodzins's (1957) work on neighbourhood dynamics of racial segregation, and related applications in sociology/ urban planning (Wolf, 1963) and economics (Schelling, 1971). Social scientists developed related concepts without explicitly using 'tipping point' terminology, such as Granovetter's work on thresholds in collective behaviour (Granovetter, 1978).

A pivotal moment, arguably, came in 2008 with a more detailed consideration of 'tipping elements' in climate system research through Lenton *et al.*'s work, which defined these as subsystems that can experience abrupt change. More recent examples of applied research on climate impacts and tipping points include Juhola *et al.* (2022) on the social conditions related to system transitions where a small change in the underlying elements or behaviour of actors triggers a large, non-linear response in the social system. Extending this point, Froese *et al.* (2023) point to a need for deeper understanding of human–nature interactions and critical linkages that may serve to provide early warning indicators for cascading tipping dynamics. Graham *et al.* (2023) add to this by noting the potential for human agency to drive transformative change while considering the equitable distribution of climate change impacts.

Pursuing a more complete discussion of the dynamics of climate thresholds, tipping points and impacts is beyond the purpose of this work. What is pertinent is consideration of what institutions might do to better apply related knowledge gained over many years. An issue considered here is how institutions and governments should respond to the challenges of climate-related tipping points and ongoing uncertainties deriving from the deep complexity of the polycrisis landscape.

Notwithstanding the breadth of research carried out to date – the details of which are only sampled in this work – questions of relevance should go beyond considering how magnificently complicated the issues are to also address what we might do to proactively and reactively address them.

## Complexity Matching Complexity: The Importance of Requisite Variety (In Teams)

An underpinning argument in this chapter is that the problems we face today, and their likely evolution, require new ways of thinking and collaborating, and arguably, different frames by which to define the operational capabilities and capacities to match future crisis landscapes.

A key premise in this work is that crises, catastrophic climate and weather disruptions and other surprises emerge from 'a background noise of regularity'. Institutional planners and decision makers, whether working on climate change adaption or disaster risk reduction, require a frame of reference that allows a suitable reduction of complexity exhibited across socio-technical-environmental systems to a scale that, while simpler, retains a semblance of real-world dynamics.

Where to begin? As a start we can look back to an element of systems thinking published some 70 years ago derived from cybernetic operations. The Law of Requisite Variety, formulated by British cybernetician W. Ross Ashby in 1954, posits that for a system to remain stable and effectively manage its environment, the variety (or complexity) of the system's responses must be at least equal to the variety of the challenges it faces (Ashby, 2013/1960). This principle is foundational in cybernetics and has applications across various fields, including biology, organisational theory and information systems.

Ashby's Law was originally framed in the context of his interest in self-regulating biological systems, but through time it has been applied to other types of systems. The British cybernetician and operational research practitioner Stafford Beer, for example, used it as the basis for his concept of viable organisational designs (Beer, 1995). Beer elaborated on the original definitions that the variety (complexity) of a control system must match the variety of the system being controlled. In other words, an organisation – or society – must have a sufficient variety of responses and strategies to cope with the variety of challenges and disturbances it faces from its environment (Beer, 1995).

In colloquial terms, Ashby's Law has come to be understood as a simple proposition: if a (managerial/oversight) system needs to deal successfully with the diversity of challenges that its (operating) environment produces, then it needs to have a repertoire of responses which is (at least) as nuanced as the problems thrown up by the environment. So, a viable system is one that can handle the variability of its environment. Or, as Ashby himself put it, 'only variety can absorb variety' (Naughton, 2017). In the context of addressing climate change and natural hazards, this suggests that the complexity and diversity of our problem-solving approaches must match the complexity and diversity of the challenges we face (Espinosa and Walker, 2017).

My argument in this chapter is that, given the emergent complexity inherent in climate change effects and tipping points and, more broadly, the emergence of polycrises, institutional regulatory, response and recovery mechanisms – embodied in the expertise of many disciplines – need to be activated and implemented. Thus, there is a need to map an ideal variety of technical skills required to provide insight into the complexities of dealing with variable climatic issues.

Attenuating the direct and indirect impacts of climate-driven disasters requires transdisciplinary coordination of research and practice – moving beyond the silos of traditional disciplined expertise – to integrate social sciences, local knowledge and governance considerations into anticipatory forms of risk governance covering the complexity of climate change (Mach *et al.*, 2024). Pohl and Hirsch Hadorn (2008) suggest that multidisciplinarity involves bringing together experts from different disciplines to work on a common problem, while transdisciplinarity goes a step further by integrating and synthesising knowledge across disciplinary boundaries to create new, holistic understanding and solutions.

Such a collaborative approach fosters a more comprehensive understanding of vulnerabilities and adaptive capacities within affected populations (Metzger *et al.*, 2004). By engaging diverse perspectives, researchers can develop more contextually relevant solutions that address not only environmental challenges but also social dimensions such as health disparities and economic inequalities. Pursuing both multidisciplinarity and transdisciplinarity can lead to several benefits. Multidisciplinarity draws on knowledge from different disciplines but stays within their boundaries. Interdisciplinarity steps further to include analyses, synthesises and harmonises links between disciplines into a coordinated and coherent whole. Transdisciplinarity integrates the natural, social and health sciences in a humanities context, and transcends their traditional boundaries (Choi and Pak, 2006).

A logical question, then, is what such collaborative, silo-busting, cross-disciplinary activity could look like? To detail one set of options, we begin with a slice out of the polycrisis idea in the form of a crisis continuum (see Fig. 1.1). Research indicates that climate change can trigger a series of interconnected effects, leading to compounded and cascading impacts that may not be immediately apparent (Pescaroli and Alexander, 2016).

Figure 1.1 represents a crisis flow from a precondition/triggering process leading to an emergent threat that creates a disruption, then to secondary and tertiary effects through time. For example, heavy rainfall can lead to flooding, which disrupts transportation networks and affects food supply chains, and waste treatment plants and potable water supplies cascading on to disrupt supply chains, exacerbating food insecurity and public health issues. With ongoing rainfall, one could see feedback loops with super-saturated soil not able to absorb additional rainfall and thus exacerbating the likelihood of more flooding leading to further disruption. Flooding is one emergent threat category that might be analysed using Fig. 1.1.

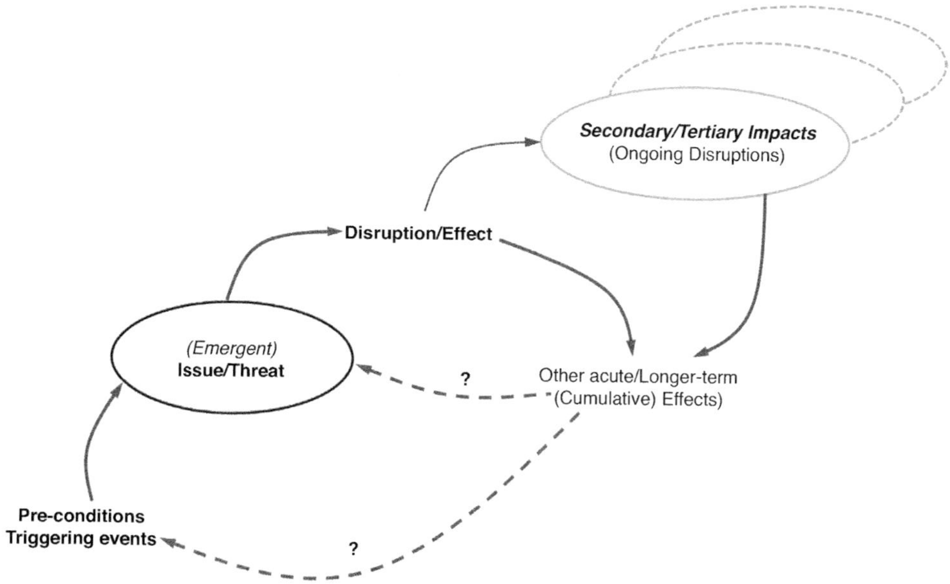


**Fig. 1.1.** A crisis continuum. Derived from Barnes, 2018; Bergin *et al.*, 2019.

Progressing further, Table 1.1 depicts an analysis of the stages of the continuum and tasks and coordinated actions across the continuum in more depth.

Multidisciplinary teams that could be active across the stages detailed in Table 2.1 should typically integrate experts from social sciences, natural sciences, technology and policy domains. These teams could include:

- systems analysts: modelling complex adaptive systems, mapping interconnections and emergence of impacts and vulnerability;
- social scientists: supporting experiential, behavioural and institutional dynamics;
- natural scientists: providing domain expertise across specific pre- and post-impact categories and scale;
- policy and legislative experts: defining policy, legislative and regulatory frameworks;
- technology specialists: implementing wide-area geospatial monitoring, communications and response systems;
- communication and engagement experts: facilitating factual knowledge transfer a well as hazard and risk communication.

In aggregate, successful multi- and transdisciplinary teams need to employ several key integration mechanisms (Alves *et al.*, 2007; Pohl and Hirsch Hadorn, 2008; Ge *et al.*, 2021; Peek and Guikema, 2021; Zebisch *et al.*, 2022):

- Knowledge and method integration: Teams need to develop shared conceptual frameworks that allow different disciplines to contribute their insights while maintaining a coherent overall approach. This would involve creating common vocabularies and establishing shared methodological approaches.
- Decision-making processes: Teams should typically design and implement structured decision-making processes that can accommodate different disciplinary perspectives while maintaining efficiency in crisis situations. This might include formal protocols for anticipating counterintuitive impacts, data integration, weighing evidence from different fields and resolving potential conflicts in interpretation and synthesis of results.
- Operational integration: Teams establish clear protocols for how different disciplinary perspectives feed into operational decisions, often using structured frameworks that allow for

**Table 1.1.** Stages and actions across the 'crisis continuum'.

| Stages | (Possible) Action |
|---|---|
| 1: Preconditions/triggering events<br>• This initial stage represents the foundational circumstances and catalytic factors that set a potential crisis in motion<br>• Characterised by the presence of underlying vulnerabilities, systemic weaknesses or emergent risk factors that create susceptibility to disruption<br>• May include both structural conditions and immediate precipitating events that function as crisis triggers | The foundational stage of crisis development encompasses both structural vulnerabilities and precipitating incidents that create conditions conducive to systemic disruption. This phase is characterised by the presence of latent risk factors, institutional weaknesses and emergent stressors that, while not immediately catastrophic, establish the underlying architecture for potential crisis manifestation. Preconditions may exist across multiple domains simultaneously, including technological, social, economic and environmental spheres, with triggering events serving as catalytic mechanisms that activate or accelerate the progression towards crisis conditions. |
| 2: (Emergent) Issue/threat<br>• Marks the transition from latent risk to active manifestation of problematic conditions<br>• Features the initial detection or emergence of anomalous patterns or concerning developments<br>• Distinguished by early warning signals that may indicate impending system stress or potential breakdown | This transitional phase marks the initial manifestation of systemic stress, and disturbances where pre-existing (and previously unknown) vulnerabilities materialise as enablers of observable perturbations within affected systems. The emergence of identifiable issues or disturbances represents a critical juncture in crisis emergence (tipping point), characterised by deviations from normal operating parameters.<br>During this stage, subtle shifts in environmental indicators may signal impending challenges, though the full scope and implications of these changes may not yet be fully apparent to stakeholders and decision makers. |
| 3: Disruption/effect<br>• Represents the primary manifestation of the crisis event<br>• Characterised by direct, first-order effects on systems, infrastructure or societal functions<br>• Marks the point where normal operating conditions are substantially altered or compromised | The disruption/effect phase represents the primary manifestation of crisis conditions, marked by substantial deviations from normal operating parameters and the emergence of clear, measurable system perturbations. This stage is characterised by direct, first-order effects that can fundamentally alter established patterns of functionality across affected systemic domains.<br>The immediate impacts may manifest as operational disruptions, damage to building stocks and the environment, resource constraints or systemic failures that require immediate intervention and adaptation strategies. This phase is likely to catalyse broader systemic changes and 'trigger' cascading effects across interconnected networks enabled by system dependencies and interdependencies. |

*Continued*

**Table 1.1.** Continued

| Stages | (Possible) Action |
|---|---|
| 4: Secondary/tertiary impacts<br>• Encompasses the cascade of subsequent effects that emanate from the initial disruption<br>• Demonstrates how impacts propagate through interconnected systems and networks<br>• Reveals indirect consequences that may emerge over time across different sectors, domains or regions | This stage encompasses the proliferation of consequential effects emanating from the primary disruption, demonstrating the interconnected nature of complex systems and their vulnerabilities. Secondary and tertiary impacts represent the propagation of crisis effects through various organisational, social and technical networks, often manifesting in unexpected domains or sectors. These derivative impacts may emerge gradually over time, presenting new challenges that require adaptive responses and potentially transforming the original crisis into a more complex, multi-dimensional challenge requiring integrated intervention strategies. |
| 5: Ongoing disruptions<br>• Represents the persistent or chronic phase of crisis impacts<br>• Characterised by long-term adaptations and systemic changes in response to initial crisis conditions and disturbances<br>• May include residual challenges, structural modifications and enduring alterations to pre-crisis norms, functions and policy settings | This additional stage in the continuum represents the long-term implications and persistent alterations to system functionality that emerge from the chain of crisis events. This phase is characterised by enduring changes to operational parameters, institutional structures and behavioural patterns that may become permanently embedded within affected systems and regions.<br>The ongoing nature of these disruptions may necessitate fundamental adaptations in organisational strategies, policy frameworks and resource allocation mechanisms – potentially leading to the establishment of new operational paradigms that reflect post-crisis realities and learned responses to experienced vulnerabilities. |

rapid but comprehensive assessment of situations from multiple perspectives and with the integration of immediate operational needs with longer-term strategic considerations.

Arguably, what is needed is not (just) more and better science, but a different application of science, characterised by deep interdisciplinarity; meaningful engagement of diverse publics; vigilant structural and systems-informed analysis of vulnerability; creative thinking that envisions multiple pathways and solutions; and downscaled analyses at regional, local and even hyperlocal levels (Rigg and Mason, 2018).

## Stepping Forward: Options for Betterment

In the vernacular there is a sporting metaphor that seems relevant to this chapter: 'skate to where the puck will be'. My interpretation of this is that anticipation is central to defining risk governance options applied to reducing the impacts of disruptions – directly or indirectly – triggered by climate variability and the almost ubiquitous driver of uncertainty that is global warming.

While considerable steps to promoting adaptation and mitigation strategies against climate disruptions are evident through past decades, the complexity of unchecked cascading impacts (polycrisis phenomena) requires different thinking. There is a need to ensure an effective understanding of the nature of this complexity of course, but stepping further requires being able to more comprehensively understand the nexus of tightly coupled dynamic systems that converge to create emergent disturbances: an anticipatory science not a reactive one.

Then with an understanding of how multi- and transdisciplinary teamwork might be put into motion, sustained effort is needed to move past the limitations of compartmentalisation of knowledge, expertise and experience to engage on intellectual and practical terms with the emergent properties and the interdependencies through and by which emergent disturbances in socio-technical-environmental systems propagate (Schweizer and Juhola, 2024).

Such insight has arguably been underrepresented in recent studies that include limited accounting for complex interactions of threats and cascading impacts and restricted coverage of regions, communities, industries and critical infrastructure at risk of extreme disturbances (Shimizu and Clark, 2015; Kwamie *et al.*, 2024). There is also a need for participatory processes that engage diverse stakeholders in knowledge co-production and deliberative decision making to navigate the uncertainties and ambiguities inherent in delivering systemic risk governance (Hochrainer-Stigler *et al.*, 2024).

A prepared future requires an active present. We need to match the vagaries and complexities of a crisis-prone world with a requisite variety of socio-technical expertise, inclusive governance and an analytical repertoire that is better suited to the evolving complexity of the global polycrisis landscape.

## References

Alves, J., Marques, M.J., Saur, I. and Marques, P. (2007) Creativity and innovation through multidisciplinary and multisectoral cooperation. *Creativity and Innovation Management* 16(1), 27–34. DOI: 10.1111/j.1467-8691.2007.00417.x.

Ashby, W.R. (2013/1960) *Design for a Brain: The Origin of Adaptive Behaviour*. Springer Science & Business Media, Berlin, Germany.

Bammer, G. (2013) *Disciplining Interdisciplinarity: Integration and Implementation Sciences for Researching Complex Real-World Problems*. ANU Press, Canberra.

Barnes, P. (2018) *Training Materials from the Risk and Resilience Program*. The Australian Strategic Policy Institute, Canberra.

Beer, S. (1995) *Brain of the Firm*, 2nd edn. Wiley, Oxford, UK.

Bergin, A., Brewster, D., Gemenne, F. and Barnes, P. (2019) *Environmental Security in the Eastern Indian Ocean, Antarctica, and the Southern Ocean: A Risk Mapping Approach*. National Security College, ANU, Canberra and the Institut de Relations Internationales et Stratégiques (IRIS). Paris. Available at: https://nsc.crawford.anu.edu.au/department-news/14413/environmental-security-eastern-indian-ocean-antarctica-and-southern-ocean-risk (accessed 29 December 2024).

Calaprice, A. (2011) *The Ultimate Quotable Einstein*. Princeton University Press, New Jersey.

Choi, B.C. and Pak, A.W. (2006) Multidisciplinarity, interdisciplinarity and transdisciplinarity in health research, services, education and policy: 1. definitions, objectives, and evidence of effectiveness. *Clinical and Investigative Medicine* 29(6), 351–364.

Drezner, D. (2023) Are we headed toward a 'polycrisis'? The buzzword of the moment explained. Available at: https:// www.vox.com/23572710/polycrisis-davos-history-climate-russia-ukraineinflation (accessed 29 December 2024).

Espinosa, A. and Walker, J. (2017) Introducing complexity and sustainability. In: *A Complexity Approach to Sustainability Theory and Application*, 2nd edn. Imperial College Press, London, pp. 1–49.

Froese, R., Andrino, A., Giudice, R., Stuch, B., Salas, S.K, *et al.* (2023) Describing complex interactions of social-ecological systems for tipping point assessments: An analytical framework. *Frontiers in Climate* 5, 1145942. DOI: 10.3389/fclim.2023.1145942.

Ge, Y., Zobel, C.W., Murray-Tuite, P., Nateghi, R. and Wang, H. (2021) Building an interdisciplinary team for disaster response research: A data-driven approach. *Risk Analysis* 41(7), 1145–1151. DOI: 10.1111/risa.13280.

Graham, S., Wary, M., Calcagni, F., Cisneros, M., Luca, C. *et al.* (2023) An interdisciplinary framework for navigating social–climatic tipping points. *People and Nature* 5, 1445–1456. DOI: 10.1002/pan3.10516.

Granovetter, M. (1978) Threshold models of collective behavior. *American Journal of Sociology* 83, 1420–1443.

Grodzins, M. (1957) Metropolitan segregation. *Scientific American Magazine* 197(4), 33. DOI: 10.1038/scientificamerican1057-33.

Hoadley, J.C. (1884) A tilting water meter for purposes of experiment. *Journal of the Franklin Institute* 117, 273–278.

Hochrainer-Stigler, S., Crimeen, A., Green, L., Moeller, H., Conaty, S.J. *et al.* (2024) Managing systemic risk through transformative change: Combining systemic risk analysis with knowledge co-production. *One Earth* 7(5), 771–781.

Huntington, H.P., Goodstein, E. and Euskirchen, E. (2012) Towards a tipping point in responding to change: Rising costs, fewer options for Arctic and global societies. *AMBIO* 41, 66–74. DOI: 10.1007/s13280-011-0226-5.

IPCC (2022) Climate Change 2022: Impacts, Adaptation and Vulnerability. Cambridge, UK: Cambridge University Press. Available at: https://www.ipcc.ch/report/ar6/wg2/ (accessed 29 December 2024).

Juhola, S., Filatova, T., Hochrainer-Stigler, S., Mechler, R., Scheffran, J. *et al.* (2022) Social tipping points and adaptation limits in the context of systemic risk: Concepts, models and governance. *Frontiers in Climate* 4, 1009234. DOI: 10.3389/fclim.2022.1009234.

Kluth, A. (2023) So we're in a polycrisis. Is that even a thing? *The Washington Post*, 2023. Available at: https://www.bloomberg.com/opinion/articles/2023-01-21/new-word-alert-polycrisis-is-being-inflicted-on-us-by-the-davos-influentials (accessed 13 July 2025).

Kopp, R.E., Shwom, R.L., Wagner, G. and Yuan, J. (2016) Tipping elements and climate-economic shocks: Pathways toward integrated assessment. *Earth's Future* Available at(8), 346–372. Available at: https://agupubs.onlinelibrary.wiley.com/doi/10.1002/2016EF000362

Kriegler, E., Hall, J.W., Held, H., Dawson, R. and Schellnhuber, H.J. (2009) Imprecise probability assessment of tipping points in the climate system. *Proceedings of the National Academy of Sciences* 106(13), 5041–5046. Available at: https://www.pnas.org/doi/10.1073/pnas.0809117106

Kwamie, A., Causevic, S., Tomson, G., Sie, A., Sauerborn, R. *et al.* (2024) Prepared for the polycrisis? The need for complexity science and systems thinking to address global and national evidence gaps. *BMJ Global Health* 9, e014887. DOI: 10.1136/%20bmjgh-2023-014887.

Lawrence, M. (2024a) Polycrisis in the anthropocene: An invitation to contributions and debates. *Global Sustainability* 7, e5. DOI: 10.1017/sus.2024.2.

Lawrence, M., Shipman, M. and Homer-Dixon, T. (2024b) *Introduction to Polycrisis Analysis: A Guide to the Cascade Institute's Approach*. Available at: https://cascadeinstitute.org/technical-paper/introduction-to-polycrisis-analysis-guide/ (accessed 29 December 2024).

Lenton, T.M., Held, H., Kriegler, E., Hall, J.W., Lucht, W. *et al.* (2008) Tipping elements in the earth's climate system. *Proceedings of the National Academy of Sciences* 105(6), 1786–1793.

Lenton, T.M., Rockström, J., Gaffney, O., Rahmstorf, S., Richardson, K. *et al.* (2019) Climate tipping points – too risky to bet against. *Nature* 575(7784), 592–595.

Levermann, A., Bamber, J.L., Drijfhout, S., Ganopolski, A., Haeberli, W. *et al.* (2012) Potential climatic transitions with profound impact on Europe. *Climatic Change* 110(3), 845–878.

Mach, K.J., Jagannathan, K., Shi, L., Turek-Hankins, L.L., Arnold, J.R. *et al.* (2024) Research to confront climate change complexity: Intersectionality, integration, and innovative governance. *Earth's Future* 12(6), e2023EF004392. DOI: 10.1029/2023EF004392.

Mark, S., Holder, S., Hoyer, D., Schoonover, R. and Aldrich, D.P. (2024) Understanding polycrisis: Definitions, applications, and responses. *SSRN Electronic Journal*. DOI: 10.2139/ssrn.4593383.

Metzger, M.J., Leemans, R. and Schröter, D. (2004) A multidisciplinary multi-scale framework for assessing vulnerability to global change. In: *The Millennium Ecosystem Assessment conference: Bridging Scales and Epistemologies, 17–20 March*, Alexandria, Egypt. Available at: https://www.millenniumassessment.org/documents/bridging/papers/metzger.marc.pdf (accessed 29 December 2024).

Milkoreit, M., Hodbod, J., Baggio, J., Benessaiah, K., Calderón-Contreras, R. *et al.* (2018) Defining tipping points for social-ecological systems scholarship – an interdisciplinary literature review. *Environmental Research Letters* 13, 033005. DOI: 10.1088/1748-9326/aaaa75.

Mocatta, G., Saville, S., Payne, N., Jansson, L., Lai, J. *et al.* (2024) Are journalists reporting the 'global polycrisis'? A report for internews. *Earth Journalism Network*. DOI: 10.26187/565t-ja43.

Mora, C., Spirandelli, D., Franklin, E.C., Lynham, J., Kantar, M.B *et al.* (2018) Broad threat to humanity from cumulative climate hazards intensified by greenhouse gas emissions. *Nature Climate Change* 8(12), 1062–1071.

Naughton, J. (2017) Ashby's law of requisite variety. What scientific term or concept ought to be more widely known? Available at: https://www.edge.org/response-detail/27150 (accessed 29 December 2024).

Peek, L. and Guikema, S. (2021) Interdisciplinary theory, methods, and approaches for hazards and disaster research: An introduction to the special issue. *Risk Analysis* 41(7), 1047–1058. DOI: 10.1111/risa.13777.

Perrow, C. (1999) *Normal Accidents: Living with High-Risk Technologies*. Princeton University Press, New Jersey.

Pescaroli, G. and Alexander, D. (2016) Critical infrastructure, panarchies and the vulnerability paths of cascading disasters. *Natural Hazards* 82, 175–192. DOI: 10.1007/s11069-016-2186-3.

Pohl, C. and Hirsch Hadorn, G. (2008) Methodological challenges of transdisciplinary research. *Natures Sciences Sociétés* 16(2), 111–121. DOI: 10.1051/nss:2008035.

Poincaré, H. (1885) Sur l'équilibre d'une masse fluide animée d'un mouvement de rotation. *Acta Mathematica* 7, 259–380.

Raymond, C., Horton, R.M., Zscheischler, J., Martius, O., AghaKouchak, A. *et al*. (2020) Understanding and managing connected extreme events. *Nature Climate Change* 10, 611–621. DOI: 10.1038/s41558-020-0790-4.

Reyers, B., Folke, C., Moore, M.-L., Biggs, R. and Galaz, V. (2018) Social-ecological systems insights for navigating the dynamics of the Anthropocene. *Annual Review of Environmental Resources* 43, 267–289.

Rigg, J. and Mason, L.R. (2018) Five dimensions of climate science reductionism. *Nature Climate Change* 8, 1030–1032. DOI: 10.1038/s41558-018-0352-1.

Schelling, T.C. (1971) Dynamic models of segregation. *The Journal of Mathematical Sociology* 1(2), 143–186. DOI: 10.1080/0022250X.1971.9989794.

Schellnhuber, H.J., Rahmstorf, S. and Winkelmann, R. (2016) Why the right climate target was agreed in Paris. *Nature Climate Change* 6(7), 649–653. DOI: 10.1038/nclimate3013.

Schweizer, P.J. and Juhola, S. (2024) Navigating systemic risks: Governance of and for systemic risks. *Global Sustainability* 7, e38. DOI: 10.1017/sus.2024.30.

Shimizu, M. and Clark, A.L. (2015) Interconnected risks, cascading disasters and disaster management policy: a gap analysis. *Planet@Risk* 3(2), 260–270. Available at: https://www.preventionweb.net/files/46564_interconnectedriskscascadingdisaste.pdf (accessed 29 December 2024).

Steffen, W., Rockström, J., Richardson, K., Lenton, T.M., Folke, C. *et al*. (2018) Trajectories of the earth system in the Anthropocene. *Proceedings of the National Academy of Sciences* 115(33), 8252–8259.

Stember, M. (1991) Advancing the social sciences through the interdisciplinary enterprise. *The Social Science Journal* 28(1), 1–14. DOI: 10.1016/0362-3319(91)90040-B.

Tooze, A. (2022) Welcome to the world of the polycrisis. *Financial Times*. Available at: https://www.ft.com/content/498398e7-11b1-494b-9cd3-6d669dc3de33 (accessed 29 December 2024).

UNDRR (2019) Global Assessment Report on Disaster Risk Reduction. United Nations Office for Disaster Risk Reduction. Geneva, Switzerland. Available at: https://www.undrr.org/publication/global-assessment-report-disaster-risk-reduction-2019 (accessed 29 December 2024).

Werners, S.E., Pfenninger, S., Slobbe, E., Haasnoot, M., Kwakkel, J.H. *et al*. (2013) Thresholds, tipping and turning points for sustainability under climate change. *Current Opinion in Environmental Sustainability* 5(3–4), 334–340. Available at: https://core.ac.uk/download/572173565.pdf (accessed 29 December 2024).

Wolf, E.P. (1963) The tipping-point in racially changing neighborhoods. *Journal of the American Institute of Planners* 29(3), 217–222. DOI: 10.1080/01944366308978066.

Zebisch, M., Terzi, S., Pittore, M., Renner, K., Schneiderbauer, S. *et al*. (2022) Climate impact chains: A conceptual modelling approach for climate risk assessment in the context of adaptation planning. In: Kondrup, C., Mercogliano, P., Bosello, F., Mysiak, J. and Scoccimarro, E. *et al* (eds) *Climate Adaptation Modelling*. Springer, Cham, Switzerland. DOI: 10.1007/978-3-030-86211-4_25.

# 2 Who Decides? Choosing Adaptation Tools for Climate Risk: A Post-normal Science Problem

**Bruna De Marchi[1] and Scira Menoni[2*]**

*[1]SVT – University of Bergen, Bergen, Norway; [2]DABC – Politecnico di Milano, Milan, Italy*

**Abstract**

This chapter explores the application of the Post-Normal Science (PNS) framework to disaster risk reduction and climate change adaptation, focusing on the challenge of integrating diverse forms of knowledge in decision making under conditions of uncertainty, complexity, and high stakes. After outlining the conceptual foundations of PNS and reviewing its growing application across disciplines and policy domains, the chapter presents an in-depth case study: a long-term collaboration between researchers and the Umbria Regional Government in Italy. This 'living lab' involved the co-development of tools for post-disaster damage and loss data collection following severe floods. The experience illustrates key PNS principles in action, including the involvement of extended peer communities, the co-production of knowledge, and the acknowledgment of multiple legitimate perspectives – scientific, administrative, operational, and local. The chapter argues that such approaches are essential for building adaptive capacity and ensuring more inclusive and effective responses to climate-related risks. It concludes by affirming PNS not as a scientific method but as a problem-solving strategy for governance under uncertainty – one that fosters democratization, pluralism, and reflexivity in science-policy interfaces.

## Introduction

The study of hazards, risks and disasters has traditionally involved a wide range of disciplines, including the natural sciences, social sciences and humanities. Each discipline emphasises different aspects of these phenomena, offering diverse perspectives that are valuable for prevention, management, mitigation and adaptation. However, the challenge lies in integrating these contributions into a cohesive vision, both for scholars and for other key stakeholders, such as policy makers, planners, practitioners and the general public.

Today, the integration of diverse perspectives and the involvement of various actors is increasingly essential, as it becomes clearer that distinguishing between 'natural' and 'human-made' risks and disasters is not always straightforward. Climate change – regarded by many as the greatest threat to humanity – serves as a prime example. Once primarily addressed through the lens of atmospheric physics, it has evolved into a multidisciplinary subject, attracting experts from a wide range of fields, as reflected in the IPCC reports from the first (FAR 1990) to the sixth one (AR6 2023).[1]

*Corresponding author: scira.menoni@polimi.it

 *Climate Change and Risk Mitigation: Reducing Vulnerabilities and Enhancing Resilience* (ed. P. Barnes)
DOI: 10.1079/9781800624153.0002

The recognition that a new approach is needed for the study and management of hazards and disasters predates the prominence of climate change. Even before the issue gained widespread attention, there were significant strides in collaboration between the natural and social sciences, although an imbalance persists to this day. For many decades, the physical sciences made impressive advancements in both knowledge and technological applications,[2] while research on the human aspects remained fragmented and often in its early stages in many countries, with the partial exception of the USA. There, since the 1950s, substantial funding was allocated to the study of human behaviour in disasters – particularly those with sudden origins, such as earthquakes – viewed as a useful model for predicting responses in the event of a nuclear attack, which was a major concern at the time.

As Meigs (2022) has noted, decades of fieldwork have consistently documented the resilience of disaster-stricken communities. These communities often mobilise a wide range of internal resources to restore essential functions and re-establish a baseline of 'normality'. Long before external aid arrives, residents spontaneously adopt new roles, take on urgent tasks and contribute to the recovery effort. Simultaneously, local organisations swiftly adjust their mandates and expand their operations to address the evolving needs of the situation. Research has repeatedly shown that active engagement, agency and creativity are far more common than inaction or passive waiting for outside help (e.g. Dynes, 1974, 1988; Quarantelli, 2008). Despite these findings, competent authorities, designated officials and professional managers often operate based on misguided stereotypes, limiting their ability to recognise and utilise local knowledge and resources that could enhance disaster response effectiveness and efficiency. Clarke and Chess (2008) coined the term 'elite panic' to describe this phenomenon, which is not limited to policy makers and senior officials but can also be seen among those overseeing field-level disaster management operations. Despite repeated warnings from social researchers, the persistent belief that the public must be shielded not only from immediate danger but also from its perceived incompetence and irrationality often leads to a patronising approach, one that can backfire and cause further harm.

This was evident in the 2011 L'Aquila earthquake, when concerns over the spread of inaccurate rumours prompted officials to devise and implement a communication strategy that, in itself, was flawed and, ultimately, counterproductive (Benessia and De Marchi, 2017). The L'Aquila earthquake highlights, among other lessons, the crucial role of the relationship between decision makers and scientific advisers, who, in this instance, were exclusively experts in earth sciences and engineering. When uncertainties are high, it is vital to incorporate a wider array of perspectives, including insights from the social and behavioural sciences. Additionally, local knowledge and traditions must be considered, as they can profoundly shape individual and collective attitudes and behaviours based on past experiences.

This chapter will show the added value that the post-normal science (PNS) approach elaborated by Funtowicz and Ravetz some thirty years ago can bring to the solution of the complex problems that have been briefly sketched above. This chapter is aimed first at illustrating the essentials of the PNS approach, second at showing the widespread fields of its application and finally at discussing an example of our own application of the PNS in the context of disaster risk reduction and adaptation to climate change impacts.

The discourse unfolds as shown in Fig. 2.1. First the PNS approach is defined and its main

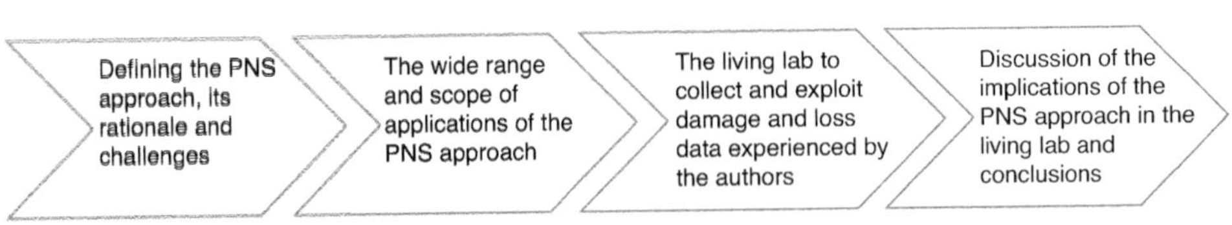


**Fig. 2.1.** The logic chain of the chapter. Author's own image.

features discussed in the second section. In the third section, the most recent applications of the PNS are described. In the fourth section, an activity carried out by the authors in strict collaboration with the Umbria Regional Government in Italy following severe floods that affected Central Italy are illustrated. The implications of the PNS approach are discussed in the fifth section. The chapter concludes with a summary of the main concepts and practical examples discussed in the chapter and possible pathways for future research and practice.

## The PNS Approach in a Nutshell

We contend that understanding the expectations and behaviours of individuals and groups – including those of the 'elites' – is essential not only for organising effective rescue and relief operations but also for designing and implementing policies aimed at preparedness and adaptation. In fact, misperceptions and incorrect expectations are widespread, often significantly hindering planning and management efforts (De Marchi and Scolobig, 2012). Research has consistently shown that integrating diverse sources of knowledge is crucial to ensure that various types of hazards, exposures and vulnerabilities are properly considered across different scales.[3] Questions arise on how we can effectively integrate insights from behavioural, social and historical research with findings from the natural sciences. How can we ensure that we don't overlook the valuable contributions of indigenous (or lay) knowledge, which, though not always articulated in strictly scientific terms, can play a crucial role in reducing vulnerability, supporting adaptation and enhancing resilience?

As will be elaborated in the third section, despite their differences, most contemporary risk issues share common features when it comes to developing effective strategies for preparedness, planning, adaptation and resilience. It is now widely recognised that nearly all contemporary risks – and even hazards – have an anthropogenic component. Given their complex and multifaceted nature, these challenges cannot be understood through a purely scientific lens, nor can they be resolved with solely technical solutions.

## The added value of the PNS to deal with complex, wicked problems at the interface between humanity and the environment

Approximately fifty years ago, Rittel and Webber introduced the concept of 'wicked problems' in the context of social policy planning. They argued that such problems cannot be clearly defined or solved through traditional scientific methods, that is, those used to address what they termed 'tame problems' (Rittel and Webber, 1973). A year earlier, nuclear physicist Alvin Weinberg had coined the term 'trans-science' to describe issues that can be framed scientifically but cannot be resolved by scientific means (Weinberg, 1972).

Building on the legacies of Rittel and Webber, as well as Kuhn (1962), Funtowicz and Ravetz developed the concept of PNS to address policy-relevant problems. These problems are characterised by four interlinked features: uncertain facts, values in dispute, high stakes and urgent decisions – the so-called PNS mantra (Funtowicz and Ravetz, 1993). Unlike simple or complicated problems, which can be framed in technical terms, complex problems defy reductionist models of analysis and action, which typically involve: (i) establishing clear facts about the issue; (ii) designing technical solutions; and (iii) implementing policies based on the previous steps. In the case of complex problems – such as climate change and many other contemporary challenges – science cannot provide uncontested knowledge or offer clear-cut solutions. The PNS diagram in Fig. 2.2 encapsulates three distinct problem-solving scenarios, each requiring different strategies. First introduced in Funtowicz and Ravetz's (1993) article, the diagram has since become iconic, reproduced by an expanding number of researchers across diverse fields.

To tackle problems within the third zone, the expertise of natural science specialists alone is insufficient, even when supplemented by that of accredited professionals. When (either or both) uncertainty and decision stakes are high, the decision-making process must incorporate

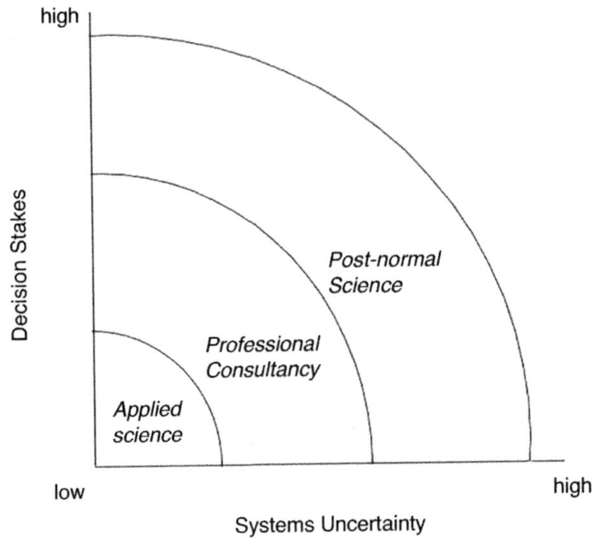


**Fig. 2.2.** PNS diagram. Funtowicz and Ravetz, 1993, with permission from Elsevier.

input from the social sciences, humanities and lay actors – including stakeholders, engaged citizens and community scientists. These individuals bring their interests, needs, values and beliefs into the discussion, contributing valuable knowledge derived from personal experience, oral traditions, indigenous practices and more. Together, they form an 'extended peer community' (Funtowicz and Ravetz, 1993, pp. 752–754), which is not only multi- and interdisciplinary but truly transdisciplinary, encompassing knowledge and expertise that are not rigidly codified.

## Challenges of the PNS approach

When the complexity of problems makes it difficult, if not impossible, to establish a single, definitive truth, efforts must shift towards the quality of the decision-making process, grounded in principles of inclusiveness, equity and fairness. As illustrated in the fourth section, participants' specific competencies are neither overlooked nor diminished; instead, they are enhanced through a process of co-generating knowledge and integrating practices that provide a more nuanced and comprehensive understanding of the issue. This also facilitates the identification – and even the imagining – of viable strategies to address it. Of course, this 'revolution' is far from easy. Scientists must overcome the 'tunnel vision' (Fjelland, 2016) imposed by their disciplines and remain open to diverse voices and perspectives. Similarly, all participants in the extended peer community must be willing to collaborate and share, free from preconceived biases. This represents a huge challenge, requiring not only the willingness to question one's own certainties but also the courage to navigate a sea of noise, confusion and contradictions, where legitimate perspectives, needs and insights are entangled with self-deception, misinformation, fake news and unreasonable claims.

We argue that the starting point in this process must be an act of belief – or, perhaps more accurately, a wager – that, against the odds, valuable, high-quality knowledge can emerge from an inclusive dialogue based on mutual openness, if not yet full reciprocal trust. In other words, conflict should be recognised as both inevitable and healthy, fostering compromise on what is desirable and achievable, rather than escalating into confrontation over incompatible, irreconcilable goals. Indeed, rather than focusing on the pursuit of a single truth, PNS emphasises the quality of the process in both research and policy. It also asserts that we must openly acknowledge that some problems may have no definitive solution and be willing to address them collectively, accepting that we may need to live with them. The last point resonates with the distinction between mitigation and adaptation with respect to climate change risk. Even in the best possible scenario of meeting the targets of emissions reduction, which unfortunately is not the case, the impacts of climate change would still be felt substantially around the globe, making adaptation crucial and unpostponable.[4]

## Recent Applications of PNS

In a 2011 editorial for a special issue on PNS, it was noted that 'Post-Normal Science (PNS) has received much attention in recent years, but like many iconic concepts, it has attracted differing conceptualisations, applications, and implications' (Turnpenny *et al.*, 2011). Indeed, PNS has gained increasing prominence across a variety of research fields related to policy regulation and the implementation of practical measures, where uncertainty is recognised as an irreducible element and new strategies for its management need to be devised.

The following are a selection of works from various fields, though this list only scratches the surface of the broader landscape: agriculture (Batie, 2008), biodiversity conservation and ethics (Boudreau LeBlanc, 2023), conservation science and sustainability (O'Connor, 2002; Bremer and Funtowicz, 2015; Giampietro and Bukkens, 2015; Guimarães Pereira and Curvelo, 2015; Buschke *et al.*, 2019), disasters such as floods and earthquakes (Amani Fard *et al.*, 2023; Park *et al.*, 2025), ecological economics (Silva and Teixeira, 2011; Strand, 2017; Martínez Alier, 2024), education (Hiis Hauge and Barwell, 2017), energy and sustainability (Giampietro *et al.*, 2013), engineering ethics (Verrax, 2017; Børsen *et al.*, 2024), food and nutrition (Kaiser and Algers, 2016; Kaiser *et al.*, 2021), health and safety (Waltner-Toews *et al.*, 2020; Gignac *et al.*, 2021; Froeling *et al.*, 2023), innovation and AI (Strand *et al.*, 2018), marine systems and fisheries (Dankel *et al.*, 2012; Stephenson *et al.*, 2016; Klimašauskaitė *et al.*, 2024), mathematical modelling and sensitivity analysis (Saltelli *et al.*, 2020), nuclear risks and waste disposal (O'Connor and Van den Hove, 2001; Eckhardt *et al.*, 2023), water cultures and management (Kovacic, 2017; Meisch *et al.*, 2022). Many contributions have also been made in languages other than English. For example, a collection of essays by Italian scholars and practitioners highlights their PNS-inspired research and policy activities across various fields (L'Astorina and Mangia, 2022).

With regard to climate-related risks, renowned climate expert Stephen Schneider stated in 2006 that 'The climate change debate – particularly its policy components – falls clearly into the postnormal science characterization and will likely remain there for decades' (Schneider, 2006, p. 608). In a review article addressing questions about the role of probabilities in climate adaptation, Dessai and Hulme write: 'This review has also showed that determining the probability of climate change cannot be resolved within what Funtowicz and Ravetz (1993) call "normal" science' (Dessai and Hulme, 2004, p. 120). Similarly, van der Sluijs argues that climate change is difficult to address using normal scientific procedures, which are primarily designed to establish clear, definitive facts. He contends that from a PNS perspective, deep uncertainty and irreducible ignorance must be acknowledged, advocating for an approach he describes as 'working deliberatively within imperfections' (van der Sluijs, 2012, p. 186). Saloranta (2001) interprets the IPCC Second Assessment Report, particularly the account of Working Group I (IPCC, 1996), as an embodiment of the extended peer community. He argues that this is the case not only because of the diverse range of actors involved in its compilation, but also because technical, methodological and epistemological uncertainties are explicitly recognised and communicated. However, writing a few years later, Beck takes a radically different stance, criticising the IPCC's linear model of expertise and calling for a broader discussion of alternatives (Beck, 2011). While Beck does not reference PNS directly, there are notable similarities in his argument, suggesting that concerns of this nature resonate across various fields.

Also, in relation to climate-related problems, it is worth mentioning two pioneering projects funded by the European Commission (EC) under the Fourth Framework Programme,[5] which applied an innovative methodology for the time. With the aid of information and communication technology (ICT) or pre-designed future scenarios, experts from different disciplines engaged in discussions on climate change with focus groups of 'lay citizens' in different European cities. In Italy, Venice was chosen for both projects: *The Ulysses Voyage* (1996–1998) (Gough *et al.*, 2003) and *Visions* (1998–2000) (Guimarães Pereira *et al.*, 2001). Much more recently, the idea of extended peer community was applied in two projects that focused on tacit forms of knowledge. One of these, *CALENDARS*, explored seasonal cultures of institutions and how they influence adaptability to seasonal variability and change. This project was conducted in the city of Bergen (Norway) and in the rural Coromandel Peninsula (New Zealand), where researchers investigated seasonality in schools and various ecological restoration and conservation efforts, including 'coastcare' groups and the traditional knowledge of Māori Iwi regarding the seasons (Meisch *et al.*, 2022).

The authors mentioned so far are not 'naïve PNS enthusiasts'. While they recognise PNS as an effective analytical model for addressing contemporary complex problems, many emphasise the challenges associated with its practical implementation, particularly regarding the concept of the extended peer community. Of course, strong criticisms also exist. For instance, Weingart dismisses PNS as 'old wine in new bottles' (Weingart, 1997), while Friedrichs argues that 'the inherent ambiguities of post-normal science have plunged climate science into a deep legitimacy crisis' (Friedrichs, 2011, p. 469).

Beyond the academic literature, PNS is frequently referenced in policy documents that analyse and discuss the features and components of advice deemed effective for assessing and addressing the uncertainties of many contemporary challenges. For example, an EC staff document notes that 'individual perceptions, judgments, and behaviours, including those of scientists, are shaped by "biases", values, and identities ... The goal therefore cannot simply be "add more science" ... The concepts of "wicked" policy problems from 1973 as well as post-normal science were developed to warn against over-reliance on technocratic solutions to policymaking and ensure trust in both science and government' (EC, 2022, p. 30).

In a report addressing the challenges of 'governance in complexity', the European Environment Agency (EEA) largely relies on PNS, suggesting it as 'one possible approach that can contribute to a paradigm shift in environmental governance' (EEA (European Environmental Agency), 2024, p. 14). In 2023, the Independent Group of Scientists appointed by the UN Secretary-General adopted the PNS framework to characterise 'issue-driven science', where 'extended peer communities can play a role in maintaining the quality and verification of knowledge' (UN, 2023, p. 90). Citing Funtowicz and Ravetz (1993) among others, the authors of a report of SAPEA (Science Advice for Policy by European Academies), in the context of the Science Advice Mechanism of the European Commission, state that 'the modern model needs to be replaced by a model of pluralistic knowledge production, aiming at enhancement of the quality and relevance of knowledge for policy, while fully acknowledging pluralism of relevant views of reality, complexity, and incompleteness of our understanding (working deliberatively within imperfections)' (SAPEA (Science Advice for Policy by European Academies), 2019, p. 84). Finally, Sir Peter Gluckman, both as Chief Science Advisor to the New Zealand Prime Minister and as Chair of INGSA (International Network for Government Science Advice), has repeatedly affirmed the value of the PNS approach in addressing contemporary issues (Gluckman, 2014, 2016).

Reading through these documents, one finds many recurrent words and concepts, including complexity, uncertainty, anticipation, precaution, participation, inclusiveness, trustworthiness, transparency, responsibility, ethics and democracy. Most of these documents also highlight climate change as one of the most challenging contemporary (policy) problems.

## The Umbria Region, Italy, Living Lab on the Co-development of Tools to Manage Post-Disaster Damage and Loss Data

In the following, a direct venture that we carried out with a regional government in Italy will be discussed. It focuses on the establishment of new procedures and tools for post-disaster damage assessment with a view to provide better understanding of the causes of damage and therefore better solutions for the future. The venture combines a technical and scientific aspect with one of policy decisions about how to carry out damage estimates and about what to consider damage. It is not by chance that in the past decade significant efforts have been made by international and European organisations into improving tools and instruments to get reliable figures for damage and losses (see for example the post-disaster needs assessment (PDNA) (GFDRR , 2013) and the activity of the Disaster Risk Management Knowledge Centre[6]). In the latter effort, the multiple use cases that enhanced post-disaster damage data collection may serve have been investigated and are currently being developed under the Risk Data Hub initiative (De Groeve *et al.*, 2013, 2014; EU Expert Working Group on Disaster Damage and Loss Data, 2015; Do O Pinto Albo *et al.*, 2018). Such multiple use cases include the forensic investigation of disasters (as conceptualised by Oliver-Smith *et al.*, 2016), accounting, necessary to programme expenses that will be needed in the future for similar types of events, and improvement and better calibration of pre-event risk models.

The basic common aspect to all those use cases is the understanding of damage and loss data as a fundamental source of knowledge that can support a variety of uses and most importantly guide towards better recovery and reconstruction, having learnt the lessons that the damage has unveiled (Menoni *et al.*, 2017). It is therefore crucial for responding to the Sendai target of 'building back better' and to better adapt systems to future impacts of climate change (UNISDR, 2015). Damage and loss estimates are produced mainly during the recovery, the phase that is stretched between the end of an emergency, when rescue and response organisations leave the scene, and the reconstruction, when both human and financial resources are fully deployed. It is a delicate phase, when needs are still pressing, when the urgency to return to normalcy is felt even more than during the emergency, as time passes from the initial event (Galderisi *et al.*, 2022). But it is also a phase when decisions, often strategic ones for prevention and adaptation, are made in the midst of a still disordered situation, on the future of places and communities.

## The Umbria Region Living Lab

The Umbria Region Living Lab constituted a unique space for sharing and co-developing knowledge of different types in a complex post-flood setting aimed at producing more consistent and reliable damage and loss data than is often the case to support recovery (Concilio *et al.*, 2022). The Umbria region is located in Central Italy (Fig. 2.3) and stretches along the Apennines, with a population of roughly

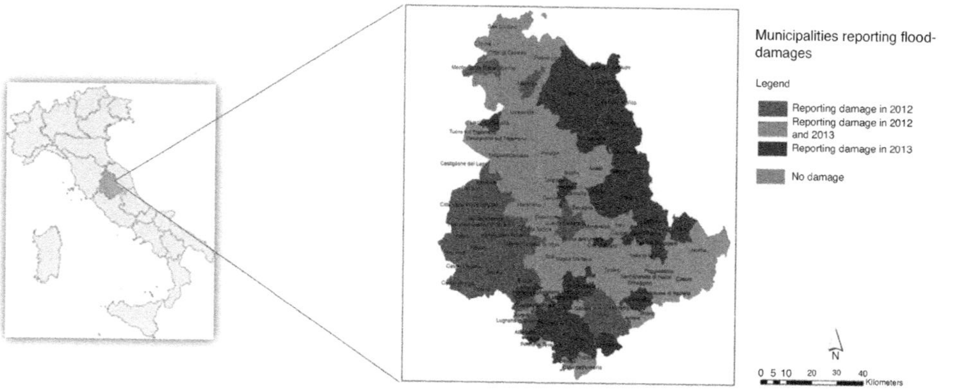


**Fig. 2.3.** Location of the Umbria region in Italy and municipalities that were affected in the two reported flood events in 2012 and 2013. Author's own image.

900,000 people who live in small towns most of which are cultural and artistic landmarks, such as Todi, Orvieto and Assisi. The region is exposed to multiple natural hazards, including seismic risk, flash and riverine floods and landslides. By being mainly a mountain region, it is exposed to both riverine and flash floods.

The Living Lab was promoted through the joint collaboration of a group of researchers of the Politecnico di Milano and the Umbria Region Civil Protection Authority. Initially, the collaboration was aimed at developing reliable models for flood risk assessments contextualised to the geographical and geomorphological setting of the region and to the specific characteristics of its built environment. However, good quality damage and loss data from previous events that are necessary for this type of calibration of flood models were missing. The director of the Weather and Hydrologic Forecasting Services of the Regional Civil Protection Authority understood the challenge and the need to set up an operational system for collecting in a structured manner post-disaster damage data to improve flood risk modelling quality and reliability.

In November 2012, a severe flood affected Central Italy, impacting the regions of Tuscany, Umbria and Lazio, with extended damage to households, agriculture, industrial and commercial activities and essential networks. The director of the Forecasting Services of the Regional Civil Protection asked the group of researchers of the Politecnico di Milano to assist the authority in post-disaster damage data collection. The request implied not only visiting the area a week after the event, but also developing new tools and new methodologies to address the large variety of exposed assets and systems that suffered damage. Whilst the technical and scientific results of this activity have been discussed in various book chapters and articles (Menoni *et al.*, 2016, 2017), the process of knowledge formation and the co-production of tools and understanding on damage will be detailed hereafter.

## The experience in a nutshell

The deployment in November 2012 actually consisted in a number of surveys that were conducted in different localities and on various types of assets, ranging from residential buildings to factories, services and infrastructures. The survey was carried out using forms that have been developed *ad hoc* and which will be described in the following section on 'tools'. The first deployment was conducted over a working week. From November 2012 to March 2015, the group of the Politecnico di Milano visited the Umbria region at least 12 times. Those visits had multiple aims. First, some of the assets that had been surveyed in November 2012 were visited again after a year to gather information on aspects such as the time that had been necessary for recovery and reopening of services and businesses at full capacity. Others were conducted to interact with the officials of the Umbria Region Civil Protection Authority and meet with local stakeholders to address in more detail damage to specific sectors, such as agriculture or essential networks. After a year, in November 2013, a new survey had to be conducted due to a new event involving flash floods and debris flows in the Northern Tiber River Catchment, in areas that had not been affected by the previous year's flood (see Fig. 2.3). The new event provided the opportunity not only to collect further data in the region but also to compare the two events and improve the overall understanding of damage, its causes and drivers.

The multiple visits were also occasions for disseminating the results of the work conducted jointly, in international seminars hosted by the Umbria region, in local high schools and in civil protection fairs with the participation of civil protection authorities from the national level and from other regions.

## Developed tools and their application

Tools for gathering damage and loss data following disasters are still rather patchy and far less systematic than required. In the context of the work conducted both within the Working Group established by the Disaster Risk Management Knowledge Centre (DRMKC) of the European Commission's Joint Research Centre (EC JRC) and in the joint activities with the Umbria region, the need to rethink and standardise such tools emerged in an evident way. First, there was the need to develop data collection tools specifically designed for a flood type of impact while most tools that already existed in Italy had been designed for seismic events. Second, most tools that existed at the time had been designed to assess damage to dwellings. Some tools were available for cultural heritage, whilst only embryonic survey forms were available for lifelines and economic activities. In the fieldwork that was conducted in the Umbria region, it became quite evident that a multi-sector damage data collection tool was needed, able to capture the multiple spatial and temporal dimensions of damage. In other words, some types of damage, for example systemic, becomes apparent at scales larger than local: failures in the transportation or power systems affect much larger areas than the single segment or point where physical damage has occurred. As for the time scale, some physical damage may be delayed (due to humidity or mold for example), and in particular, systemic indirect damage due for example to business interruption can be estimated only months after the impact.

The tools needed for such multi-scale and multi-sector damage and loss data collection are also key for the various use cases for which this data can serve.

Operationally, new forms were developed for different sectors, in particular for economic and commercial activities, taking into consideration the need also to capture damage to equipment, machinery and both rough material and finished products. The survey forms were developed initially as drafts and then refined and improved following application in the field, testing exercises and discussion between researchers and regional officers. Considering that in the longer term regional officers, municipal officers and volunteers would be required to fill in the survey forms, the latter were progressively standardised and rendered more similar to forms that are already in use by surveyors for damage caused by earthquakes. In fact, the co-development of such tools was made in an effort to tailor them as much as possible to the capacities and the competences available within the public administration and the volunteers (Berni *et al.*, 2017). Training has also been organised with both, having in mind the need to anchor the introduction of the new tools to the existing procedures and keep them in line with already familiar types of work and instruments.

## Framing Co-production of Knowledge and Tools in the Umbria Region Case within PNS

A number of aspects of the venture presented above embed PNS as a reference mode of producing knowledge and solutions to complex problems raised in the context of disaster risk and climate change.

### Responding to a real need

First, the activity started in response to a real need that was recognised by the officials of the Umbria Regional Civil Protection Authority in preparing for and preventing floods. Consequently, funding mechanisms were sought to allow for the procedure of damage and loss data collection and analysis to mature. At the beginning, the Umbria region contracted the Politecnico di Milano working group using part of the Structural Funds that are allocated by the EC to regions in Europe to develop projects and reforms to guarantee the overcoming of persisting economic and social gaps. Disaster risk reduction and climate change adaptation are general goals that fit well with European objectives and are therefore not only eligible but even encouraged as types of measures to be funded. Subsequently, the effort was financed through four projects funded by the European Commission: one under the 7th Framework Programme (FP7), one under Horizon 2020[7] and two by the Directorate-General for Civil Protection and Humanitarian Aid Operations (DG ECHO). Each funding allocation not only enabled the venture to continue but also provided the opportunity to address the topic from a slightly different angle, exploring a rather wide range of technical and procedural aspects, including experimenting with forms of knowledge co-creation.

### Acknowledging the legitimacy of multiple forms of knowledge

Coherently with the PNS approach, the Enabling Knowledge for Disaster Risk Reduction and

Climate Change Adaptation (Know4DRR)[8] project was developed on the assumption that there are different forms of knowledge that must be mobilised in disaster risk reduction and climate change adaptation: not only the scientific knowledge, but also other forms, namely legislative, organisational and common knowledge. All of them have been relevant in the accomplishment of the Umbria venture.

Scientific knowledge was key in eliciting the indicators and data that are required for a full and cross-sectoral damage analysis and for fully unleashing their potential in multiple use case applications. The organisational knowledge provided by the public administration representatives and by the stakeholders of different sectors, from industry to essential services, was necessary to provide guidance on how things are actually done and to select what may work and what may not and for what reasons. The legislative knowledge was relevant to provide insight into existing procedures within which the new methodology and instruments for collecting damage data could be mainstreamed and to understand what eventually could be proposed for future norms and rules. Local knowledge was generally provided by the people with whom the group of researchers and civil protection officials interacted during the surveys. They provided significant insight on prior exposure to hazards in the area where their activity or home was located. They were also very interested in gaining better understanding of their own exposure and vulnerabilities and on how improved land use and urban development or redevelopment could support enhanced prevention for the future.

The roles and types of knowledge are not a fixed 'datum' but can actually change and evolve over time. We used the term 'pracademic' (Duncan *et al.*, 2014) to indicate those who were actually able to cross, and more than once, the boundary between practitioners and researchers. Some of the officials of the Civil Protection Authority had a PhD and had been researchers on hydrology in the Italian National Research Council (CNR). Researchers of the Politecnico di Milano worked closely with civil protection officers, visiting the affected sites and jointly collecting the data. The officers also contributed actively to some scientific publications, bringing operational insight on the uncertainties implied, for example, in early warnings.

### Involving multiple stakeholders in the Umbria Living Lab

The Umbria Region Living Lab has been successful in developing innovation, not only in terms of the products that have been delivered in the form of an advanced information system for collecting, analysing and sharing damage and loss data, but also in terms of the processes through which such innovation has been achieved. An important outcome has been setting the premises of a shared operational procedure between the group of researchers of the Politecnico di Milano and the Umbria Civil Protection Authority, demonstrating how different steps of the damage and loss data collection and analysis could be performed jointly, assigning tasks that can be accomplished partly on location and partly remotely. The resulting procedure could be adopted in principle also by other research institutions interested in getting precious data while supporting public administrations in the delicate phase of recovery (Berni *et al.*, 2017).

The intense relationship that developed over the course of 2 years and several visits permitted the involvement of a large number of local and regional stakeholders and the co-development of an understanding of damage to multiple sectors in the immediate aftermath of the two events, in the short and medium terms. Damage and loss data are far less uncontestable and objective than usually believed. As correctly put by Handmer (2003), there is no such a thing as objective damage assessment, as different stakeholders weigh differently damage to different sectors. This is not only because of personal interests, but often genuinely because of the different perspective actors have on the impact and of their responsibility in the emergency and recovery management (Molinari *et al.*, 2014). The fact that Umbria is a small region helped in gathering around the table all lifeline providers, in the water, power, telecommunications and transportation sectors, as well as mayors and representatives of local authorities from many of the affected municipalities. Particularly interesting for the type of insights that we have been able to gain was the meeting with the manager of an oasis that hosts rare ecosystems of animal and vegetation species, who shared detailed information of the damage that was suffered by the infrastructures of the oasis as well as by the various ecosystems. He also highlighted the administrative challenges in dealing with a compensation system that does not foresee a wide range of impacts beyond the more frequently addressed ones, such as damage to households or roads.

### A mutual learning experience for all

Overall, the experience has taught very relevant lessons to each of the participants. Officers gained important understanding on how damage and loss data could serve a variety of purposes, not just to fill the administrative 'bureaucratic' requirement for getting compensation of losses. Officers got access to a large set of information regarding practices adopted in other regions and countries and had a feeling of how their experience could contribute to the wider endeavor that was taking place in Europe and internationally. The multi-scale nature of damages has also been better understood by all participants to the Living Lab. In fact, damages due to disasters involve larger areas and more sectors than the one directly affected, given the systemic interconnections in supply chains, economic activities and services. On the one hand, the very local impact of such events was discussed with the communities and the victims themselves, who were reporting not only what damage and losses they had incurred but also, inevitably, how the event had impacted on their lives. On the other hand, some actors, including local stakeholders, were able to participate in meetings of the European-funded projects, thus getting the opportunity to meet their peers in other countries and compare their experience of extreme events with that of others.

Getting to the higher European level also gave access to instruments that were either unknown or underestimated in their potential benefit for local authorities and regions. For example, the full potential of the Copernicus service[9] has become better known thanks to the cooperative activity and requests that were advanced for the first time by the Region in the aftermath of the 2013 flood and debris flow event. As researchers, we became familiar with the public administration procedures and became better able to tailor the information system that we designed to the actual needs and to the iterative requests coming from the central government for compensation purposes. We were able to appreciate that each of the use cases for which the information system was developed had its own challenges and constraints, even those that we imagined to be relatively straightforward, like compensation. The latter, in fact, evolves over time due to multiple iterations responding to what can be actually granted in terms of funding allocation or compensated by the insurance system. The engineers in our group understood that the over-refinement of flood models to assess in the most precise way variables such as discharge and water depth is not always useful for practical purposes. Evaluating variables such as the presence of sediments and pollutants that are rarely considered is instead far more important in assessing potential damage to structures.

### Conclusion

Our aim was to demonstrate the usefulness of the PNS approach in research and practice in the field of disaster risk reduction and climate change adaptation. We provided an overview of the PNS framework, referencing authors who have applied it across various fields and topics. We then presented a living lab in which a group of researchers worked with a regional public administration to co-develop tools for improved post-disaster damage and loss data collection and analysis. Improved damage and loss data are also necessary to identify trends over past decades and to verify the claim of the increasing influence of climate change as a causal factor of growing levels of damage (Pielke, 2014). In fact, in order to be able to demonstrate the causation link between climate change and increased levels of damage, better data on damage are needed as, with the quality of existing data, the attribution of increased damage to climate change is not possible.

We have shown that the effort conducted across academic and practitioner communities involved the uptake of different forms of knowledge beyond just scientific knowledge. The many applications of the PNS described in the section 'Recent Applications of PNS' and the

Umbria Region Living Lab discussed in the two sections that followed provide tangible evidence of the many forms that the application of this approach may take and the variety of contexts in which it can be profitably applied.

This is coherent with the description of PNS provided by its founders, who insist that it is not a new scientific method or paradigm. Instead, it is to be understood as 'a problem-solving strategy for policy-related research' in contexts where it is impossible to separate facts from values. PNS seeks to provide an interpretation of the co-evolution between science and society, while also aiming to serve as a movement for the democratisation of science and expertise (Funtowicz and Ravetz, 2023, p. 435).

## Notes

[1] https://en.wikipedia.org/wiki/Intergovernmental_Panel_on_Climate_Change#Assessment_reports: FAR, SAR, TAR, AR4, AR5, AR6. The 2023 date for AR6 is that of the synthesis report of those published in 2021 and 2022.

[2] See for example https://euobserver.com/stakeholders/157390

[3] https://extranet.who.int/kobe_centre/sites/default/files/pdf/WHO%20Guidance_Research%20Methods_Health-EDRM_3.2.pdf

[4] https://climate.nasa.gov/solutions/adaptation-mitigation/

[5] Framework Programmes (FP) are set by the European Commission to fund research in different forms, individual and collaborative, aimed at science investigation and application and innovation development.

[6] https://drmkc.jrc.ec.europa.eu/

[7] These are multiannual research programmes that are designed and managed by the European Commission. FP7 ran between 2007 and 2013, whilst Horizon 2020 ran between 2014 and 2020.

[8] The project titled Enabling Knowledge for Disaster Risk Reduction and Climate Change Adaptation was funded under the FP7 and run between 2015-2017.

The project titled Enabling Knowledge for Disaster Risk Reduction and Climate Change Adaptation was funded under the FP7 and run between 2015-2017.

[9] The European Earth Observation Programme Copernicus is implemented in partnership between a number of European agencies and member states and provides a wide range of satellite-based services, including early warning and rapid post-disaster damage assessment.

## References

Amani Fard, F., Riekkinen, K. and Pellikka, H. (2023) Evaluating the effectiveness of land-use policies in preventing the risk of coastal flooding: Coastal regions of Helsinki and Espoo. *Land* 12, 1631. DOI: 10.3390/land12081631.

Batie, S. (2008) Wicked problems and applied economics. *American Journal of Agricultural Economics* 1176–1191. DOI: 10.1111/j.1467-8276.2008.01202.x.

Beck, S. (2011) Moving beyond the linear model of expertise? IPCC and the test of adaptation. *Regional Environmental Change* 11, 297–306. DOI: 10.1007/s10113-010-0136-2.

Benessia, A. and De Marchi, B. (2017) When the earth shakes … and science with it. the management and communication of uncertainty in the L'Aquila earthquake. *Futures* 91, 35–45. DOI: 10.1016/j.futures.2016.11.011.

Berni, N., Molinari, D., Ballio, F., Minucci, G. and Munoz, C.A. (2017) Best practice of data collection at the local scale: the RISPOSTA procedure. In: Molinari, D., Ballio, F. and Menoni, S. (eds) *Flood Damage Survey and Assessment: New Insights from Research and Practice*. Wiley, New York, pp. 77–94.

Boudreau LeBlanc, A. (2023) Building the bioethics tools of a community council to the future: The ecosystemic gap. *Humanities and Social Sciences Communication* 10, 562. DOI: 10.1057/s41599-023-02038-6.

Bremer, S. and Funtowicz, S. (2015) Negotiating a place for sustainability science: Narratives from the Waikaraka Estuary in New Zealand. *Environmental Science & Policy* 53, 47–59. DOI: 10.1016/j.envsci.2014.11.006.

Buschke, F.T., Botts, E.A. and Sinclair, S.P. (2019) Post-normal conservation science fills the space between research, policy, and implementation. *Conservation Science and Practice* 1(8), e73. DOI: 10.1111/csp2.73.

Børsen, T., Chance, S. and Meskens, G. (2024) A post-normal environment-centered approach to engineering ethics education. In: *The Routledge International Handbook of Engineering Ethics Education*. Routledge, Abingdon, UK, pp. 108–124.

Clarke, L. and Chess, C. (2008) Elites and panic: More to fear than fear itself. *Social Forces* 87, 993–1014. Available at: https://www.jstor.org/stable/20430900

Concilio, G., Menoni, S. and Norton, J. (2022) Innovation environments and risk management: Results from the KNOW4 European project. In: De Lucia, C., Borri, D., Kubursi, A. and Khakee, A. (eds) *Economics and Engineering of Unpredictable Events: Modelling, Planning and Policies*. Routledge, Abingdon, UK, pp. 200–215.

Dankel, D.J., Aps, R., Padda, G., Röckmann, C., Sluijs, J.P. *et al.* (2012) Advice under uncertainty in the marine system. *ICES Journal of Marine Science* 69, 3–7.

De Groeve, T., Poljansek, K. and Ehrlich, D. (2013) *Recording Disaster Losses: Recommendation for a European Approach. JRC Scientific and Policy Report*. Publications Office of the European Union, Luxembourg. Available at: https://publications.jrc.ec.europa.eu/repository/bitstream/JRC83743/lbna26111enn.pdf (accessed 14 July 2025).

De Groeve, T., Poljansek, K., Ehrlich, D. and Corbane, C. (2014) *Disaster Loss Data Recording in EU Member States. A Comprehensive Overview of Current Practice in EU Member States*. JRC Scientific and Policy Report.

De Marchi, B. and Scolobig, A. (2012) The views of experts and residents on social vulnerability to flash floods in an Alpine region of Italy. *Disasters* 36(2), 316–337. DOI: 10.1111/j.1467-7717.2011.01252.x.

Dessai, S. and Hulme, M. (2004) Does climate adaptation policy need probabilities. *Climate Policy* 4, 107–128. DOI: 10.1080/14693062.2004.9685515.

Do O Pinto Albo, A., Marin Ferrer, M., Poljansek, K. and Casajus Valles, A. (2018) *Disaster Damages and Loss Data for Policy: Pre- and Post-Event Damage Assessment and Collection of Data for Evidence-Based Policies*. EUR 29080 EN. Publications Office of the European Union, Luxembourg.

Duncan, C., Scherer, S. and Wade Apicella, S. (2014) HFA thematic review: Research area 2, priority for action 3, core indicator 3. Relevant information on disasters is available and accessible at all levels, to all stakeholders (through networks, development of information sharing systems etc.). Background paper prepared for the 2015 Global Assessment Report on Disaster Risk Reduction. Available at: http://www.preventionweb.net/english/hyogo/gar/2015/en/bgdocs/UNISDR,%202014c.pdf (accessed 14 July 2025).

Dynes, R.R. (1974) *Organised Behaviour in Disasters*. Ohio State University Press, Columbus, Ohio.

Dynes, R.R. (1988) Cross-cultural international research: Sociology of disasters. *International Journal of Mass Emergencies and Disasters* 6(2), 101–129.

EC (European Commission) (2022) *Supporting and Connecting Policymaking in the Member States with Scientific Research*. Commission Staff Working Document SWD(2022) 346 Final. European Commission, Brussels. Available at: https://knowledge4policy.ec.europa.eu/sites/default/files/SWD_2022_346_final.PDF (accessed 7 July 2025).

Eckhardt, A., Krütli, P. and Röhlig, K.-J. (2023) Extended peer reviews for resilient safety. *Safety of Nuclear Waste Disposal* 2, 83–84. DOI: 10.5194/sand-2-83-2023.

EEA (European Environmental Agency) (2024) *Governance in Complexity. Sustainability Governance under Highly Uncertain and Complex Conditions*. EEA Report 05/2024. Available at: http://www.yrdcpcn.com/upload/2024/0702/55faada3-85a9-4d4e-8578-af06e3724104.pdf (accessed 14 July 2025).

EU Expert Working Group on Disaster Damage and Loss Data (2015) *Guidance for Recording and Sharing Disaster Damage and Loss Data: Towards the Development of Operational Indicators to Translate the Sendai Framework into Action*. JRC Scientific and Policy Report. Publications Office of the European Union, Luxembourg.

Fjelland, R. (2016) When laypeople are right and experts are wrong: Lessons from love canal. *HYLE–International Journal for Philosophy of Chemistry* 22(1), 105–125.

Friedrichs, J. (2011) Peak energy and climate change: The double bind of post-normal science. *Futures* 43(4), 469–477. DOI: 10.1016/j.futures.2010.12.004.

Froeling, F., Gignac, F., Toran, R., Ortiz, R., Ficorilli, A. *et al*. (2023) Implementing co-created citizen science in five environmental epidemiological studies in the CitieS-Health project. *Environmental Research* 240, 2. DOI: 10.1016/j.envres.2023.117469.

Funtowicz, S.O. and Ravetz, J.R. (1993) Science for the post-normal age. *Futures* 25(7), 739–755. DOI: 10.1016/0016-3287(93)90022-L.

Funtowicz, S. and Ravetz, J.R. (2023) Post-normal science. In: Padilla Rosa, E. and Ramos-Martín, J. (eds) *Elgar Encyclopedia of Ecological Economics*. Edward Elgar Publishing, Cheltenham, UK, pp. 435–437.

Galderisi, A., Menoni, S., Setti, G. and Tognon, A. (2022) Disaster recovery reform and resilience. In: Eslamian, S. and Eslamian, F. (eds) *Disaster Risk Reduction for Resilience: Disaster and Social Aspects*. Springer International Publishing, Cham, Switzerland, pp. 25–54.

GFDRR (Global Facility for Disaster Reduction and Recovery) (2013) *Post Disaster Needs Assessment: Volumes A and B*. GFDRR, Washigton, DC. Available at: https://www.gfdrr.org/en/publication/post-disaster-needs-assessments-guidelines-volume-2013 (accessed 7 July 2025).

Giampietro, M., Mayumi, K. and Şorman, A. (2013) *Energy Analysis for a Sustainable Future. Multi-Scale Integrated Analysis of Societal and Ecosystem Metabolism*. Routledge, Abingdon, UK.

Gignac, F., Froeling, F.E.M., Hoek, G., Vermeulen, R., Nieuwenhuijsen, M *et al*. (2021) Narrative review of citizen science in environmental epidemiology: Setting the stage for co-created research projects in environmental epidemiology. *Environment International* 152, 106470. DOI: 10.1016/j.envint.2021.106470.

Gluckman, P. (2014) Policy: The art of science advice to government. *Nature* 507(7491), 163–165. DOI: 10.1038/507163a.

Gluckman, P. (2016) The multiple dimensions and challenges of the science-policy interface. Second International Conference on Government Science Advice, Brussels, 29 September 2016. Available at: https://ingsa.org/wp-content/uploads/2023/01/INGSA-opening-PDF.pdf (accessed 17 July 2025).

Gough, C., Darier, E., Marchi, B., Funtowicz, S., Grove-White, R. *et al*. (2003) Framing the experience: Contexts of citizen participation in IA. In: Kasemir, B., Jäger, J., Jaeger, C. and Gardner, M.T. (eds) *Public Participation in Sustainability Science: A Handbook*. Cambridge University Press, Cambridge, UK, pp. 37–61.

Guimarães Pereira, Å. and Curvelo, P. (2015) In the name of sustainability. *International Journal of Sustainable Development* 18(4), 247–260. DOI: 10.1504/IJSD.2015.073695.

Guimarães Pereira, Å., Corral Quintana, S., De Marchi, B., Funtowicz, S., Gallopin, G. *et al*. (2001) *Visions. Adventures into the Future*. European Commission Report EUR 19926 EN. EC Joint Research Centre, Ispra, Italy.

Giampietro, M. and Bukkens, S.G.F. (2015) Quality assurance of knowledge claims in governance for sustainability: Transcending the duality of passion vs. reason. *International Journal of Sustainable Development* 18(4), 282–309.

Handmer, J. (2003) The chimera of precision: Inherent uncertainties in disaster loss assessment. *The Australian Journal of Emergency Management* 18(2), 88–97.

Hiis Hauge, K. and Barwell, R. (2017) Post-normal science and mathematics education in uncertain times: Educating future citizens for extended peer communities. *Futures* 91, 25–34. DOI: 10.1016/j.futures.2016.11.013.

IPCC Houghton, J.T., Meira Filho, L.G., Callander, B.A., Harris, N. and Kattenberg, A. (eds). (1996) *Climate Change 1995 – the Science of Climate Change*. Contribution of Working Group I to the Second Assessment Report of the Intergovernmental Panel on Climate Change (IPCC). Cambridge University Press, Cambridge, UK, p. 572.

Kaiser, M. and Algers, A. (2016) Food ethics: A wide field in need of dialogue. *Food Ethics* 1(1), 1–7.

Kaiser, M., Goldson, S., Buklijas, T., Gluckman, P., Allen, K. *et al*. (2021) Towards post-pandemic sustainable and ethical food systems. *Food Ethics* 6, 4. DOI: 10.1007/s41055-020-00084-3.

Klimašauskaitė, A., Dees, P., Funtowicz, S.O., Drivdal, L.E. and Berta, O.G. (2024) Science at the Secretariat of the International Seabed Authority: Post-normal science for recalibration of policy instruments. *Open Research Europe* 4, 252. DOI: 10.12688/openreseurope.18879.1.

Kovacic, Z. (2017) Investigating science for governance through the lenses of complexity. *Futures* 4(252). DOI: 10.1016/j.futures.2017.01.007.

Kuhn, T. (1962) *The Structure of Scientific Revolutions*. University of Chicago Press, Chicago, Illinois.

L'Astorina, A. and Mangia, C. (2022) *Scienza, Politica e Società: L'approccio Post-Normale in Teoria e Nelle Pratiche*. Collana Scienziati in Affanno? Edizioni CNR (Consiglio Nazionale Delle Ricerche) Rome, Italy, DOI: 10.26324/SIA1.PNS.

Martínez Alier, J. (2024) *Land, Water, Air and Freedom. The Making of World Movements for Environmental Justice*. Edward Elgar Publishing, Cheltenham, UK.

Meigs, J.B. (2022) Elite panic vs. the resilient populace. *Commentary: The Monthly Magazine of Opinion*. Available at: https://www.commentary.org/articles/james-meigs/elite-panic-vs-the-resilient-populace/ (accessed 17 July 2025).

Meisch, S.P., Bremer, S., Young, M.T. and Funtowicz, S.O. (2022) Extended peer communities: Appraising the contributions of tacit knowledges in climate change decision-making. *Futures* 135, 102868. DOI: 10.1016/j.futures.2021.102868.

Menoni, S., Molinari, D., Ballio, F., Minucci, G., Mejri, O *et al*. (2016) Flood damage: A model for consistent, complete and multipurpose scenarios. *Natural Hazards and Earth System Sciences* 16(12), 2783–2797.

Menoni, S., Atun, F., Minucci, D., Berni, N., Molinari, D. *et al*. (2017) Defining complete post flood scenarios to support risk mitigation strategies. In: Molinari, D., Ballio, F. and Menoni, S. (eds) *Flood Damage Survey and Assessment: New Insights from Research and Practice*. Wiley, New York, pp. 151–171.

Molinari, D., Ballio, F., Handmer, J. and Menoni, S. (2014) On the modelling of significance for flood damage assessment. *International Journal of Disaster Risk Reduction* 10, 381–391.

O'Connor, M. and Van den Hove, S. (2001) Prospects for concertation on nuclear risks and technological options: Innovations in governance practices for sustainable development in the European union. *Journal of Hazardous Materials* 86, 77–99.

O'Connor, M. (2002) Social costs and sustainability. In: Bromley, D.H. and Paavola, J. (eds) *Economics, Ethics and Environmental Policy: Contested Choices*. Blackwell Publishing, Oxford, UK, and Malden, Massachusetts, pp. 181–202.

Oliver-Smith, A., Alcántara-Ayala, I., Burton, I. and Lavell, A. (2016) *Forensic investigations of disasters (FORIN): A conceptual framework and guide to research*. IRDR FORIN Publication no.2. Integrated Research on Disaster Risk, Beijing.

Park, W., Grace, M., Hutton, C. and Knowles, S.G. (2025) Science education in an age of unnatural disasters: An introduction to the special issue. *Science & Education* 34, 957–967. DOI: 10.1007/s11191-025-00652-1.

Pielke, R. (2014) *The Rightful Place of Science: Disasters and Climate Change*. Arizona State University Press, Tempe, Arizona.

Quarantelli, H.L. (2008) Conventional beliefs and counterintuitive realities. *Social Research: An International Quarterly of the Social Sciences* 75(3), 873–904.

Rittel, H. and Webber, M. (1973) Dilemmas in a general theory of planning. *Policy Sciences* 4(2), 155–169.

Saloranta, T.M. (2001) Post-normal science and the global climate change issue. *Climate Change* 50, 395–404.

Saltelli, A., Benini, L., Funtowicz, S., Giampietro, M., Kaiser, M. *et al*. (2020) The technique is never neutral. How methodological choices condition the generation of narratives for sustainability. *Environmental Science & Policy* 106, 87–98. DOI: 10.1016/j.envsci.2020.01.008.

SAPEA (Science Advice for Policy by European Academies) (2019) *Making Sense of Science for Policy under Conditions of Complexity and Uncertainty*. SAPEA, Berlin, Germany.

Schneider, S.H. (2006) Climate change: Do we know enough for policy action? *Science and Engineering Ethics* 12(4), 607–636.

Silva, M. and Teixeira, A. (2011) A bibliometric account of the evolution of EE in the last two decades: Is ecological economics (becoming) a post-normal science? *Ecological Economics* 70(5), 849–862. DOI: 10.1016/j.ecolecon.2010.11.016.

Stephenson, R.L., Stacey, P., Pastoors, M.A., Kraan, M., Holm, P. *et al*. (2016) Integrating fishers' knowledge research in science and management. *ICES Journal of Marine Science* 73(6), 1459–1465. DOI: 10.1093/icesjms/fsw025.

Strand, R. (2017) Post-normal science. In: Spash, C.L. (ed.) *Routledge Handbook of Ecological Economics*. Routledge, Abingdon, UK, pp. 288–297.

Strand, R., Saltelli, A., Giampietro, M., Rommetveit, K. and Funtowicz, S. (2018) New narratives for innovation. *Journal of Cleaner Production* 197, 1849–1853. DOI: 10.1016/j.jclepro.2016.10.194.

Turnpenny, J., Jones, M. and Lorenzoni, I. (2011) Where now for post-normal science?: A critical review of its development, definitions, and uses. *Science, Technology & Human Values* 36(3), 287–306. DOI: 10.1177/0162243910385789.

UN (2023) *Global Sustainable Development Report 2023: Times of Crisis, Times of Change: Science for Accelerating Transformations to Sustainable Development*. United Nations, New York.

UNISDR (2015) *Sendai Framework for Disaster Risk Reduction 2015-2030*. Available at: https://www.preventionweb.net/files/43291_sendaiframeworkfordrren.pdf (accessed 17 July 2025).

van der Sluijs, J.P. (2012) Uncertainty and dissent in climate risk assessment: A post-normal perspective. *Nature and Culture* 7(2), 174–195. DOI: 10.3167/nc.2012.070204.

Verrax, F. (2017) Engineering ethics and post-normal science: A French perspective. *Futures* 91, 76–79. DOI: 10.1016/j.futures.2017.01.009.

Waltner-Toews, D., Biggeri, A., Marchi, B., Funtowicz, S., Giampietro, M. *et al*. (2020) Post-normal pandemics: Why COVID-19 requires a new approach to science. Steps Centre. *Recenti Progressi in Medicina* 11(4), 202–204. DOI: 10.1701/3347.33181.

Weinberg, A. (1972) Science and trans-science. *Minerva* 10, 209–222.

Weingart, P. (1997) From “finalization” to “mode 2”: Old wine in new bottles? *Social Science Information* 36(4), 591–613. DOI: 10.1177/053901897036004002.

# 3 Ethics in Climate Risk Mitigation and Policy Development

**Mark Crosweller***
*Founder and Director – Ethical Intelligence Pty Ltd, Australia*

**Abstract**

"This chapter explores the ethical complexities facing disaster management leaders as climate-induced disasters intensify. Highlighting Australia's recent cascading events, it critiques policy shortfalls and questions the fairness of burdening citizens with disproportionate responsibility. It advocates for a renewed social contract grounded in shared responsibility, trust, and relational ethics. It also critiques leadership tendencies toward invulnerability and argues for compassionate, context-aware, and participatory governance. Ethical frameworks—virtue, deontological, utilitarian, and relational—are examined to guide decision-making. Ultimately, the chapter calls for ethical, compassionate leadership to reduce suffering and rebuild trust in the face of growing climate risk."

## Introduction

The lived experiences of many of the world's citizens have brought to our collective consciousness myriad compounding and cascading risks that have resulted from a multiplicity of climate change drivers and the complex interactions that occur between them. As the recent history of global disasters has shown, these compounding and cascading risks driven by climate change have resulted in continuous or contiguous disaster events that are producing large-scale impacts on lives and livelihoods. This chapter will explore some of the ethical challenges that disaster management leaders face when having to advise governments on policies and strategies for mitigation, preparedness, response and recovery.

## Setting the Scene

Globally, many people have experienced, and many more people have witnessed, natural hazard events influenced by the effects of climate change increasing in their intensity, frequency and impacts. These hazards-turned-disasters are also becoming increasingly complex in nature, cascading from one event to the next. In addition, they are compounding upon each other and running consecutively, leaving little time for communities or those that support them to recover. Often, the recovery time frames need to be measured in years to decades, not weeks or months (Gibbs *et al.*, 2021). Most recently, Australia has experienced an almost continuous stream of natural hazard events, from 2019 through to the present day, that reflect this trend.

*Corresponding author: mark.crosweller@ethicalintelligence.com.au


DOI: 10.1079/9781800624153.0003

These events commenced with long-term catastrophic drought conditions across much of the Australian continent. The 2019/2020 financial year was the sixth driest year on record (24% below the long-term average). Spring 2019 was Australia's driest spring on record, and December 2019 the driest December on record. Also, December 2019 was Australia's hottest ever month, 18 December 2019 was Australia's hottest ever day and 17–24 December 2019 the hottest ever week. The accumulated forest fire danger index (FFDI) for December 2019 was the highest ever recorded in December over most of the country, and the root-zone soil moisture in December 2019 was in the lowest 10% of all Decembers since 1900 (AIDR, 2020).

Amid these climatic conditions, severe to catastrophic fires erupted in all states and territories, resulting in the loss of 33 lives, over 3000 homes, more than three billion animals, and 24 million ha of land. Financial estimates of loss and damage exceeded A$10 billion (Binskin *et al.*, 2020). Soon thereafter, extensive thunderstorm activity, hailstorms and heavy rainfall leading to flooding were variously experienced in Queensland, New South Wales, the Australian Capital Territory and Victoria (often in similar areas to where the fires had impacted that same year) as well as tropical cyclones in Western Australia and the Northern Territory. COVID-19, while not a climate-influenced natural hazard, unfolded across the world in early 2020, severely hampering the ability of many communities to proficiently recover from the impacts of the fires and floods experienced in 2019/2020 (AIDR, 2020).

The 2020/2021 financial year saw the highest overnight temperatures ever recorded during spring in Australia. The mean spring temperature was slightly more than 2°C above the long-term average, and November recorded a mean temperature 2.47°C above the long-term average, making it the hottest November on record. Importantly, the Bureau of Meteorology released data indicating that the Australian annual mean temperature had increased by around 1.4°C since 1910. Between October and December, Frazer Island in Queensland lost more than half its landmass to bushfires, while during February, bushfires in the Perth Hills in Western Australia destroyed 86 homes and approximately 11,000 ha of bushland (AIDR, 2021).

During March 2021, influenced by a weak La Niña, the east coast of Australia experienced extreme rainfall that led to widespread flooding in New South Wales, affecting regions from the North Coast to the Sydney metropolitan area in the south. Several Sydney suburbs experienced their worst flooding in 60 years. Far south-eastern communities in Queensland were also affected by flooding and heavy rainfall. During this flood event, three people died and damage estimates exceeded $1 billion. In June 2021, an extensive severe weather system across the state of Victoria resulted in significant damage and widespread flash flooding, largely across the Dandenong Ranges and Gippsland areas. Over 160,000 properties lost power, many for over a week, and at least 110 properties were declared uninhabitable. Two deaths were reported, and many communities remained in communications isolation for a number of days (AIDR, 2021).

The 2021/2022 financial year saw Australia's wettest November on record (since 1900), influenced by a moderate strength La Niña, with rainfall recorded at 135% above long-term average. NSW had its wettest November on record, with more than 3 times its average November rainfall observed during the month. Nationally, in autumn 2022 the mean minimum temperature was the second highest on record (behind only 2016) (AIDR, 2022).

Between 28 October and 1 November 2021, severe thunderstorms brought torrential rainfall, destructive wind gusts and abundant hail to areas of South Australia, Victoria and Tasmania, causing power outages at 518,000 properties in Victoria and over 30,000 properties in Adelaide. The event was declared an insurance catastrophe by the Insurance Council of Australia, which recorded over 109,000 claims worth $839 million in damages. On 9 January 2022, tropical cyclone Tiffany crossed the Queensland coast and tracked initially west into the Northern Territory and then south as a significant weather system. This resulted in widespread flooding that caused extensive damage to South Australian transport infrastructure and freight routes, cutting direct rail and road freight access between Darwin, Adelaide and Perth for several weeks (AIDR, 2022).

From late February to early April, the east coast of Australia endured three intense weather systems that led to record rains and flooding, causing significant damage to critical infrastructure, property, the agricultural sector and numerous critical supply chains. The flood event was one of the most severe in Australia's history, with over 14 million people impacted in New South Wales and Queensland. Twenty-six people lost their lives and over 8000 homes were damaged or destroyed, with an estimated value of insurance claims exceeding $4.29 billion (AIDR, 2022).

The 2022/23 financial year saw the climatic combination of a negative Indian Ocean Dipole, a positive Southern Annular Mode and a La Niña event (for the third year in a row) that contributed towards wetter-than-average conditions for much of Australia. Spring was the wettest on record for New South Wales, Victoria and the Murray–Darling Basin as a whole (AIDR, 2023). In early 2022, extensive coastal areas of South-East Queensland and New South Wales, including the Northern Rivers Region and the Hawkesbury-Nepean River Catchment to the west of Sydney, suffered catastrophic levels of flooding that resulted in the loss of nine lives, over 15,000 homes impacted with over 5000 uninhabitable and estimates in excess of $4 billion of damage, not to mention the significant loss of plants, animals and other biodiversity (Fuller and O'Kane, 2022).

## Frameworks and Reviews

These events are but a small snapshot of an even more extensive picture of disaster impact caused by natural hazard events influenced by climate change that have occurred in Australia between 2019 and 2023.[1] This snapshot reveals that the impacts and resultant consequences that have been experienced and/or witnessed by many Australians are severe, extensive, complex, compounding, cascading, prolonged and increasing in intensity and frequency. It also reveals that the climate is continuing to produce record-breaking weather events that contribute to the causes of these impacts.

Importantly, and as pointed out by numerous scholars (Wisner *et al.*, 2012; National Resilience Taskforce, 2018; O'Connell *et al.*, 2018; Kelman, 2020), climate and weather are not the only causes of disaster. Where, why and how we place ourselves upon the landscape and the extent to which we consider the forces of nature (earth, wind, fire and water) when assessing for exposure and vulnerability are more significant factors than only considering the climate and its resultant weather. These assessments include land-use planning policies, building construction standards and social, economic and environmental policies targeted towards disaster risk reduction, resilience and vulnerability reduction.

These scholars argue, in part, that governments need to pay greater attention to the potential of these forces through our ability to assess, analyse and model climate and natural hazard risk, and then fully act upon that analysis with the intention of minimising impacts and consequences as much as possible (i.e. by reducing vulnerability and exposure and increasing adaptive capacities across all communities). These scholars argue that, in so doing, we might begin to see a reversal of the potential for increasing impacts and resultant loss and damage that are likely to occur as we continue to populate into hazardous areas concomitant with deteriorating climate and weather conditions.

For many people actively engaged in disaster risk management (incorporating risk reduction, resilience, prevention, preparedness, response, relief, recovery and reconstruction), this approach is not new. In the almost 40 years that I have been involved with the disaster management and national security sectors, arguments in support of such approaches have been consistently prosecuted in myriad fora, producing frameworks that attempt to address these concerns. For example, the Sendai Framework for Disaster Risk Reduction (UNISDR, 2015), the (Australian) National Disaster Risk Reduction Framework (NDRRF, 2018) and both the New Zealand and the Australian National Strategy for Disaster Resilience (NSDR-AUS, 2011; NDRS-NZ, 2019) are but a small sample of many influential frameworks that attempt to address climate and disaster risk.

Likewise, post-disaster reviews such as the Royal Commission into National Natural Disaster Arrangements (RCNNDA) (Binskin *et al.*, 2020) and the New South Wales Independent Flood

Inquiry (Fuller and O'Kane, 2022) form part of a substantial picture of reviews and inquiries that have produced in excess of 140 (independent, agency, audit, parliamentary, coronial and royal commission) reports. These reports have generated in excess of 1400 recommendations across 32 themes covering the spectrum of disaster management (Cole *et al.*, 2018) that seek to contribute towards addressing these concerns.

Such frameworks and reviews have tended to generate, at least in the short term, increases in investments for most elements of disaster management, albeit there are substantial arguments about where this investment should be directed. For example, there are many who argue that increasing investments in disaster risk reduction and resilience initiatives can yield returns of between 1:4 and 1:8 for every dollar invested in response and/or recovery (Deloitte, 2013, 2016).

Partly in reply to this argument, the Commonwealth Government established the Disaster Ready Fund (DRF) as its flagship disaster resilience and risk reduction initiative. This initiative (valued at $1 billion over 5 years) aims to deliver projects that support Australians to manage the physical and social impacts of disasters caused by climate change and other natural hazards. This funding is also being tied to the implementation of various frameworks, such as the NDRRF.

The DRF's objectives are to: (i) increase the understanding of natural hazard disaster impacts as a first step towards reducing disaster impacts in the future; (ii) increase the resilience, adaptive capacity and/or preparedness of governments, community service organisations and affected communities to minimise the potential impact of natural hazards and avert disasters; and (iii) reduce the exposure to risk, harm and/or severity of a natural hazard's impacts, including reducing the recovery burden for governments, cohorts at disproportionate disaster risk and/or affected communities.[2] State and Territory governments have also committed to facilitating projects and initiatives in support of the DRF through matched funding (cash and in-kind). Some have also committed to independent investments, such as the New South Wales Government's investment of $1.9 billion towards frontline resourcing for their emergency services (Elliott, 2021).

These frameworks and reviews have also sought to make explicit the notion of a social contract of shared responsibility for disaster risk management between governments, businesses (the private sector), non-governmental organisations, communities and individuals (McLennan and Handmer, 2014; Hunt and Eburn, 2018).

## The Social Contract of Shared Responsibility

The social contract of shared responsibility is a foundational concept in political philosophy, with influential thinkers like Thomas Hobbes, John Locke and Jean-Jacques Rousseau contributing to its development (Knowles, 2001; White, 2012). While the specifics of the social contract may vary across different political and cultural contexts, the fundamental idea remains that citizens unite to form a society, agreeing to certain responsibilities and obligations for the greater good. Citizens willingly accept certain obligations and responsibilities in exchange for the benefits and protection provided by the community (D'Agostino *et al.*, 2019).

The social contract of shared responsibility typically encompasses several key elements. Citizens agree to fulfil certain duties and responsibilities, known as mutual obligations, that contribute to the overall wellbeing of the community, including obeying laws, respecting the rights of others and actively participating in civic life. In return, citizens expect certain rights and protections from the community or government that may include personal freedoms, security and equal treatment under the law. There is also an emphasis on a commitment to the common good, whereby citizens acknowledge that their actions should contribute positively to the welfare of the entire community, not just their individual interests. A commitment to principles of justice and fairness whereby citizens agree to a system of rules and laws that are designed to be just and applied fairly to all, ensuring equal opportunities and protection, is also important. Finally, there is an understanding of reciprocity, where citizens contribute to the collective wellbeing and, in return, benefit from the contributions of others, fostering a sense of interdependence and shared responsibility (Quinn, 2017; Sobhy, 2021).

## The Risk of Invulnerability

However, shared responsibility for disaster risk management is a highly contested concept. Exactly who is responsible for what and to what extent the degree or proportion of that responsibility rests with either citizens or governments is the subject of much academic debate (Singh-Peterson *et al.*, 2015; Lukasiewicz *et al.*, 2017; Atkinson and Curnin, 2020; Crosweller and Tschakert, 2021). Research conducted in 2019 would suggest that many disaster management leaders, influenced by socio-cultural factors such as neoliberalism, tend to bias responsibility for disaster resilience towards citizens and away from governments (Crosweller and Tschakert, 2020, 2021).

For example, this bias tends to expect citizens to exercise their agency irrespective of their financial, health and other circumstances, rather than acknowledging that certain social, economic and structural supports need to be in place before citizens could exercise agency. Likewise, taking the pressure off response and recovery instead of having reasonable access to community resources, and the expectation of always being situationally aware instead of being able to access free and full information, are examples of other biases that potentially put undue pressure on citizens, and particularly those that are systemically vulnerable and disadvantaged (Crosweller and Tschakert, 2020).

This tension between disaster management governance and leadership and the citizen regarding how responsibility should be shared emerges across all levels of disaster intensity, from low through to catastrophic, but is not equitably shared. At the lower end of disaster intensities, governments, their agencies and the more privileged members of the community have the collective capacity to effectively treat risk such that there is relatively little impact in comparison to what otherwise would happen if risk treatment capabilities were not in place (Crosweller and Tschakert, 2019).

These capabilities would include things like fire appliances, helicopters, flood boats and the like, but also legislation, policy, construction standards, various forms of knowledge and intelligence, and the citizen's capacity and capability to manage their own risk, and so forth. In these circumstances, shared responsibility appears, at least on the surface, to work reasonably well. However, if the citizen experiences various dimensions of vulnerability on a daily basis, their experience is likely to be different. This often results in loss and suffering that those more privileged do not experience at these lower levels of intensity (Crosweller and Tschakert, 2019).

When disaster intensities reach severe to catastrophic levels, not only do those less privileged become vulnerable to extensive loss and suffering, but the effectiveness of most risk treatments by governments and even the most capable citizens are also exceeded. This all too often results in extensive loss and suffering across the broader community, including the government, as has been witnessed in Australia between 2019 and 2023. When disaster intensities exceed the capacity for governments, their agencies and citizens to cope, a space of mutual vulnerability is entered (Crosweller and Tschakert, 2019).

Vulnerability is a pervasive and universal feature of the human condition, experience and identity, rather than only a property of marginalised or disadvantaged groups (Beckett, 2006; Gilson, 2011, 2014). Irrespective of who we are or the circumstances in which we find ourselves during the course of our lives, we can be susceptible to harm and suffering without the ability to cope and adapt (IPCC, 2014). Even though the causes, extent, duration, intensity, frequency and specific circumstances in which we and others (such as governments and agencies) become vulnerable can be highly variable and often driven by exposure to unfavourable circumstances, we all have ethical obligations towards each other to ensure we and others are no longer susceptible to the harmfulness and suffering that arises from being vulnerable (Li, 2017). This is especially true if we hold power and influence over others (Goodin, 1985a, b).

It is also important that we contextualise vulnerability as both positive and negative in the way it is experienced. Positively, it opens opportunities for compassion, love, relationality and support. Negatively, it opens up possibilities for suffering, harmfulness and violence (McLeod,

2012). However, if we (as disaster management leaders) only view vulnerability as something to be negated, we run the risk of developing a worldview driven by fear, defensiveness, denial, avoidance and disavowal of vulnerability – that is, we risk becoming invulnerable (Gilson, 2011, 2014).

## Two Key Questions

It is this mutual space of vulnerability and the risk of becoming invulnerable that reshapes the tension of shared responsibility between governments, their agencies and the citizens to an ethical one driven by two key questions. Firstly, how can trust and confidence be upheld between the government, its agencies and the citizens they are called upon to lead, serve and protect, which is essential to the social contract of shared responsibility, when vulnerability becomes pervasive? Secondly, how can otherwise avoidable suffering and vulnerability be negated, minimised or managed effectively as a pragmatic response to the need for protection under the social contract (while recognising that some suffering and vulnerability cannot be avoided)? In my experience, both questions are existential. The answers are highly complex, and they change over time as the context of disasters changes, as for example with the ongoing deterioration of the climate concomitant with inadequate policy responses. This notwithstanding, these questions, shaped by mutual vulnerability, speak to the heart of the social contract of shared responsibility between governments and citizens. They speak to the need for relationality.

## Why Trust is Important

Trust is an essential ethic in any relationship, whether that relationship be intimate, romantic, platonic, familial, transactional, institutional and so on. At its most foundational level, to trust is to have firm belief and confidence in the reliability, truth, competence and willingness of someone or something to do something or make something happen that would otherwise open us up to being exposed, vulnerable, harmed or betrayed (Oxford English Dictionary, 2017; Faulkner, 2018). When we decide to trust someone or some entity (a collection of people), we are willingly accepting a level of risk or vulnerability. To place our trust in them, we need to have a confident expectation that the person or entity will not cause us further harm or suffering (D'Olimpio, 2018).

It is this commitment to trust that establishes the bond between government and institution and citizens in the social contract of shared responsibility (Etzioni, 1998; Bishara and Schipani, 2009; Cowden and Singh, 2017), along with the democratic processes that help us all to determine our collective understanding of the common good (Ridley-Duff, 2007). Placing such trust in governments and their institutions is, according to sociologist Anthony Giddens, ontological (Giddens, 1990, 1999). That is, it is foundational to our sense of being. As a member of a governed and functioning society, we rely upon others (governments, agencies, institutions – public and private), including their advice, expertise and expert systems (such as knowledge systems, resources and so on), for a host of protections that we as citizens are unable to provide. We need to do this as a way of easing our existential anxieties about the unpredictability of life so that we can get on with other priorities towards finding our happiness, wellness and capacity to flourish (Lupton, 2013).

We need to create environments of cognitive and behavioural certainty that are stable, predictable and fair, so that as citizens we can feel 'safe'. Placing trust in others (i.e. governments and their agencies) to afford us protection is a way of achieving this sense of safety (Mitzen, 2006; Poppo *et al.*, 2016). Because trust is interpersonal and context specific, the characteristics of trust must be present in the trustee (the person or persons within an entity to be trusted) and the trustor (the person or persons within an entity who trusts) (Terblanche and Heyns, 2020). When trust is present, harmful conflict is reduced, transaction costs are lessened, expedient relationships flourish and responses to crises become more effective (Rousseau *et al.*, 1998). When we seek to trust another person or an entity, we are seeking to bind them into a relationship and attitude of responsibility, which may also require them to alter their behaviour (Szerszynski, 1999).

### Institutional Factors that Support Trust and Reduce Suffering and Vulnerability

Recent research suggests there are several institutional factors that are needed to contribute towards the ontological security of the citizen and the trust required to sustain the social contract of shared responsibility to reduce suffering in the face of climate and disaster risk (Crosweller and Tschakert, 2019, 2020, 2021).

Firstly, governments need to ensure that the necessary structures are in place such that ideas about how to provide protection can be contested and negotiated through participative and representative democratic processes. This needs to occur both among citizens and between citizens and their governments. Such an approach can then be used to determine the balance between citizens' rights and responsibilities to achieve safety and inform decision-making processes about how to collectively manage emerging risks (Lodge, 2009). This includes the need to value processes of social engagement in policy and strategy development and decision making, understand and appreciate the complexity of human and socio-political behaviour, treat citizens with dignity and recognise citizens for their moral worth and innate rights (Caro, 2015, 2016). These factors are often found, for example, in community resilience programmes and other initiatives that directly engage communities as part of the decision-making processes about how to provide protection. However, often these programmes are not so much about shared problem definition and solution but rather about government-defined problems that require citizens to participate in government-defined solutions (McLennan and Handmer, 2012, 2014). The distinction is subtle but important.

Secondly, governments need to use their power to regulate, plan, fund, prioritise, facilitate and deliver on community safety needs. This will necessarily involve distributing the equities of power, wealth and resources made available to them in a fair and equitable manner to minimise risk. Generally, this is achieved through various disaster management programmes to reduce hazard, vulnerability and exposure, concurrent with establishing safe locations and sustainable livelihoods for citizens (Wisner *et al.*, 2012). Unfortunately, all too often this approach, when it is taken, suffers from short-termism. For example, often the funding is identified post event and tied to relatively short-term budget cycles, whereas often longer-term funding and investment are required. Policy and strategy development can also suffer from a lack of long-term vision.

Thirdly, governments need to incorporate strategies to address socio-economic causes of vulnerability, inequity and injustice directly into resilience policy frameworks. This would include putting in place the structural, social, environmental and economic frameworks to address these causes so that citizens would not be as susceptible to harm and suffering, and/or would have sufficient capacity to cope and adapt to the emerging circumstances. This would allow them to be adequately supported in their endeavours to protect themselves and others as part of a fully functional community (Crosweller and Tschakert, 2020). This factor is perhaps the most challenging of all.

The systemic causes of vulnerability stretch across most if not all areas of society and, in practical terms, it is extremely difficult to governance them effectively unless addressing causes becomes a principal tenet of the government's social, environmental and economic policy agenda. It is also difficult, but not impossible, for disaster management leaders to have sufficient influence on these policy agendas such that the root causes of systemic vulnerability are identified and addressed. Expanding post-disaster inquiry terms of reference to explore systemic dimensions of vulnerability could be one way of bringing greater focus and action to this important factor.[3]

Fourthly, governments need to ensure that economic, scientific and technical knowledge is created and made accessible to assist citizens and others in understanding their commensurable risks (Beck, 1992, 2009), as well as social and cultural knowledge to address non-commensurable risks. These risks would include things that we cannot place an economic value upon, such as a sense of place, relics, artefacts and memories (Barnett *et al.*, 2016; Tschakert *et al.*, 2017). Timely, accurate and accessible information informing situational awareness is also critical (Handmer and James, 2007).

However, one key cause of the erosion of trust between governments and citizens has been in the production of knowledge that informs their expertise in risk. The continual updating of knowledge through reflexive processes and access to alternative forms of (reliable and unreliable) knowledge through emerging technologies have exponentially increased over the past 20–30 years. This has resulted in citizens increasingly casting doubt upon the previously unquestionable expertise of public and private institutions (Giddens, 1998). Instead, citizens are relying more and more upon their own knowledge through myriad sources, official and unofficial, reliable and unreliable, and less upon governments and their agencies (Wachinger *et al.*, 2013). This and other related concerns were well explored by the RCNNDA, and numerous recommendations were made to address them (Binskin *et al.*, 2020).

For example, the RCNNDA found that there were significant inconsistencies across the nation in much of the information and data that governments and others needed to make informed decisions about when managing natural disaster risk. This included information regarding hazards, emergency warnings, air quality and situational awareness. It also included how governments and other organisations managed information about the impact of past disasters, and about how exposure and vulnerability to future hazards was collected, stored and shared. The RCNNDA called for a national approach for greater cooperation and coordination across governments and agencies; a greater sharing of resources across jurisdictions; an agile emergency response and recovery capability, with skills and technology that can be used across the country; and the data, systems and research to help manage and mitigate disaster risk efficiently and effectively (Binskin *et al.*, 2020).

When these institutional factors are present, citizens can look to the expertise and expert systems of disaster management leaders for guidance on how they can protect themselves with a sense of confidence, while avoiding unnecessary stress and anxiety and reinforcing their ontological security (Giddens, 1991; Krolikowski, 2008). This has the beneficial effect of underwriting their capacity to exercise their agency by helping them choose courses of action that have favourable outcomes while still sharing responsibility for managing risk with government in accordance with a social contract of shared responsibility (Rogers and Frederick, 2000; Bickerstaff *et al.*, 2009).

## Leadership Factors that Undermine Trust and Increase Suffering and Vulnerability

These institutional factors provide insight to some of the critical policy changes, underpinned by an ethic of trust, that are required to address the deteriorating disaster environment described at the beginning of this chapter. However, alone they are not sufficient. During my research over the past few years, I have detected a pervasive sense of invulnerability amongst disaster management leaders, that is, a propensity to deny, disavow or ignore the fact that in severe to catastrophic disaster intensities, governments, their agencies and the citizens they are asked to lead, serve and protect become mutually vulnerable (in different ways and at different scales) (Crosweller and Tschakert, 2019).

Such invulnerability is driving an insensitivity towards the suffering and vulnerability of others by us as disaster management leaders, our organisations and our governments. This insensitivity is driven not only by our inability to see the suffering of others through ignorance, denial or disavowal, but also by our propensity to be overconfident in our abilities and our tendency to act paternalistically when seeking to alleviate it. Our personal and professional ambitions can also stand in the way, as can our privileged positioning within our communities and our personal and institutional fear of being vulnerable (Crosweller, 2022a).

When we become insensitive to the genuine needs of others, we risk being insufficient in our responses to that suffering. We can lack the necessary commitment to fully appreciate and act upon the suffering of others, and we can prioritise our commitments towards cost avoidance, productivity, efficiency and effectiveness rather than addressing genuine needs of citizens. Such insensitivity and insufficiency tends to limit responses towards suffering to duty, obligation and compliance. We only do what the law and our systems, processes, frameworks and organisational cultures require us to do. We become risk-averse, and our innovation lessens, as does our compassion. Finally, we become defensive when criticised. Rather than opening up to reflexive practice, we tend to blame others, including citizens and nature (Crosweller, 2022a).

We also defend our systems, policies, frameworks, organisational cultures and the like, sometimes making incremental changes, but rarely introducing significant reform to our governance and practices (Crosweller, 2022a, b). The RCNNDA made several observations and recommendations in this regard (Binskin *et al.*, 2020). For example, it noted that 'Implementing our recommendations calls for a cohesive and unified national effort. National natural disaster arrangements are a shared responsibility. Failure by governments to act on our recommendations will shift risk to others' (p. 33), implying 'others' would be communities and citizens.

In short, when we are invulnerable towards those we are asked to lead, serve and protect, otherwise avoidable suffering and vulnerability flourishes (Crosweller, 2022a).

## Leadership Factors that Support Trust and Reduce Suffering and Vulnerability

The evidence to suggest that invulnerability is happening comes from several sources. Independent commissions of inquiry, including parliamentary and legal commissions, specific witness statements obtained during these inquiries and insights from disaster management leaders themselves provide compelling evidence. However, this same body of evidence also indicates that there is a pathway out of this invulnerability. There is an opportunity for the experience of mutual vulnerability and suffering to establish the basis of relationality between us as leaders in disaster management and those we are asked to lead, serve and protect.

Breaching the divide of invulnerability to become relatable is possible when we are open to emotional engagement and affectivity, the need for change of systems and approaches, and not feeling compelled to have to know everything up front. Being tolerant of the differences of others and being open to changing oneself are also beneficial. Also, having the capacity to 'walk in the shoes of another', develop a level of distress tolerance and genuinely appreciate our common humanity – that everyone suffers and is vulnerable – is equally important. Being non-judgemental and reflecting upon personal experiences of loss and suffering also makes a substantial difference in our ability to relate to those we are being asked to lead, serve and protect (Crosweller, 2022a).

With a relatable disposition, we are more likely to be committed to alleviating suffering and vulnerability. We are also more likely to have an affective and empathetic sense of purpose, mission and willingness to assist others, and be more motivated to act in myriad ways to alleviate suffering and vulnerability. We are also more likely to be venturous in how we do this. We increase our potential to willingly learn and improve from previous experiences as well as take more risks and pursue difficult courses of action, politically, operationally and administratively. We are more able and willing to set aside our privilege and put the needs of others ahead of our own. We are also more likely to willingly engage with citizens, colleagues and communities in decision making as well as being willing to express our own limitations. Critically, despite the risks associated with being venturous, our willingness to be held to account for our actions, driven by a strong sense of our own ethics, values and principles that are overtly expressed to others, increases. In short, by being more relatable to those we are asked to lead, serve and protect, we are more compassionate, and otherwise avoidable suffering is reduced (Crosweller, 2022a).

## The Ethical Context of Relationality

As disaster management leaders, we have varying degrees of access to the institutional equities of power, wealth and resource. We also have varying degrees of access to senior policy and decision makers who make decisions about how these equities are to be distributed. This access to power, wealth, resource and decision making helps facilitate our moral, political and legal obligations to uphold trust and confidence. It also helps to negate, minimise or manage the suffering and vulnerability of those we are called upon to lead, serve and protect (i.e. to be compassionate), all within a social contract of shared responsibility. As I have argued, exercising relationality in this context is essential. However, there is one final question we need to consider: In what normative ethical context can we best develop relationality?

As leaders, consciously or subconsciously, we make ethical decisions daily about how best to lead and serve others and protect them from suffering and vulnerability. The ethical basis on which we can make such decisions relies upon our own moral character, a sense of duty towards others and a desire to maximise happiness and minimise suffering for as many others as possible. In normative ethical theory, these approaches are known as virtue ethics, deontological ethics and utilitarian ethics.

Virtue ethics focuses on the character of ethical agents. Aristotle argued that we are acting with virtue when we respond at the right time, to the right objects, towards the right people, with the right motive and in the right way (Carr, 2003). When we act with virtue, we are acting through good habits of body, speech and mind directed towards the good. In this context, 'good' is defined as fostering the ultimate end for humans (to be happy and well), satisfying the fundamental nature of being human (to flourish) or the intrinsic goods of the virtues themselves, such as compassion, kindness and care (Dawson, 2015).

Deontological ethics focuses on duties of ethical agents. Emmanuel Kant argued that all moral decisions should be founded upon a 'categorical imperative'. That is, we should consider how we should act rather than considering only the end results. Also, people should not be used as means to an end, but rather be treated as ends in themselves (Guyer, 2002). Acting ethically in this way leads us to consider two important questions: What would the world look like if everyone confronting this dilemma acted in this way? and How would I feel if the same decision was taken that impacted upon me? Kant's categorical imperative suggests that there is a universal solution to be found to any moral dilemma (Bruton, 2004). For example, all disaster survivors should be treated with the same level of fairness, respect and dignity irrespective of their circumstances.

Utilitarian ethics focuses on the outcomes of actions of ethical agents. John Stuart Mill argued that all moral decisions should be taken upon a single criterion: securing the greatest utility (happiness) for the greatest number. This suggests that decisions should be made in terms of their overall outcomes for as many people as possible (Chryssides and Kaler, 1993). For example, the suffering that arises from disaster impacts and consequences should be minimised for as many people as possible in the most efficient and effective way.

To summarise, we can act in a virtuous way that builds good character and is consistent with a good life (virtue), we can follow prescribed duties that should in turn be for everyone else to follow (deontological) or we can maximise welfare or minimise harm (utilitarian) (Morrell and Dahlmann, 2022). All three normative ethical frameworks aim to increase happiness, wellbeing and flourishing while minimising suffering and vulnerability.

Notwithstanding, unlike virtue ethics, both deontology and utilitarianism are principles-based. Put another way, they are useful and necessary at scale and can more easily translate into governance mechanisms that can apply across entire sectors, countries or even the global population (Morrell and Dahlmann, 2022). The United Nations Sustainable Development Goals and the Sendai Framework for Risk Reduction are two key examples of frameworks underpinned by principles-based ethics.

However, principles-based ethics can be criticised for treating ethical problems through dispassionate calculation (Overman Smith and Thompson, 2002) and for being too abstract (Velasquez, 2001). For example, they can treat complete strangers with the same level of consideration as well-known friends, colleagues, community members and so on. They can also fail to recognise and address what is truly valued and important in the context of place. In so doing, they fail to take into account the contextual, contingent and relational nature of leadership which is essential to its function (Ladkin, 2006).

To demonstrate, during my appointment as Director General of Emergency Management Australia, a story was relayed to me of a community who were significantly aggrieved by the actions of emergency services, not because they acted immorally or unethically, but because they did not understand the context of what the community valued. When a large bushfire was approaching a wilderness area, the fire brigade responded to protect the local infrastructure: a tourism centre and associated facilities. The brigade was successful in protecting the structures from damage, and firefighters thought they had done (and in fact had done) a great job.

However, the local community was highly disappointed because, as they rightly pointed out, tourists did not come to look at the tourism centre, they came to visit the wilderness area and its forests. Now that the forests were destroyed, there would be no tourists for at least 5 years! The brigade had responded ethically by prioritising the principles of life, property and the environment (in descending order) but had failed to appreciate the local context of what was most valuable. Had the brigade understood this, they could have still met the principles of response, but also given greater focus and attention to what the community most valued. And to do that, the brigade would have needed to act through a 'relational ethic' with the local community.

A relational ethic means that we recognise the entwined nature of our relationship with others (Cunliffe and Eriksen, 2011) and that we both shape and are shaped by our social experiences in everyday interactions and conversations (Gergen, 1999). In so doing, we develop an understanding of what is meaningful to others, we work through differences as a moral responsibility and we are guided by the practical wisdom of our lived experiences and our good character (virtue) (Cunliffe and Eriksen, 2011; Metz and Clark Miller, 2016). By being 'in relationship' with others while thinking, speaking and acting with virtue as a moral responsibility, we are seeking to realise a more perfected understanding and practical embodiment of the good for human life (MacIntyre, 1981/2007). Through a relational ethic premised upon the virtues of trust and compassion, leaders could have established trust with the local community by recognising those things that they most valued, were vulnerable to the impact of fire and other hazards and would cause immense suffering if lost.

## The Challenge for Leaders

The social contract of shared responsibility, as we have discussed, requires both citizens and governments to honour mutual obligations, agree upon what is seen as the common good, and commit to reciprocity within an environment that respects rights and protections as well as justice and fairness. The institutional factors mentioned above that support trust and reduce suffering and vulnerability provide a pragmatic approach to how these shared commitments may be met. To deliver on these commitments, leaders need to think, speak and act relationally.

Therefore, the challenge for leaders is to navigate the tension between a set of broad and general ethical principles aimed at securing trust and reducing suffering of the general populace that are ascribed in our various frameworks and the more nuanced needs of citizens shaped by the values, norms and other cultural factors of communities. To ease this tension, there are three interrelated factors that can help refine our understanding of what is important to citizens, increase our ability for relatedness and decrease the risk of being insensitive to their needs. The three factors are sentiment, symbolism and data.

As leaders, we need to grant ourselves permission, through our ability to show compassion, to understand how people are 'feeling' about the risks they are likely to face as well as what they are 'thinking' about them. When we take this course of action, we begin to achieve a 'sense' of the commonality of feeling that becomes the community's 'sentiment'. Sentiment refers to the emotional tone or attitude expressed in our communication that reflects our feelings and opinions, which can be positive, negative or neutral, or more nuanced, such as anger, sadness, happiness and so on (Slovic, 2007).

Having gained an appreciation of the sentiment of community, we can look for the 'symbolism' – the story of one who represents the feelings of the majority. We can use the power of our imagination, through images, narrative, story and first-person testimony of the individual who suffers as a symbol of the suffering of many to allow us to connect and feel what the other is experiencing. This approach can motivate us towards an appropriate compassionate response, not only towards that one individual, but to the many others they symbolise or represent (Farmer, 2010). As one highly experienced Australian government minister expressed to me in my research:

> I think it gets harder once it goes to scale. Because things become abstract again. And the way you find compassion is through stories of individuals rather than collective groups. So, I think you have to try and relate the stories of individuals to that larger scale to keep it human and not abstract. So that's why finding and hearing stories and relating them in the context of response to a suffering of a group is important. Otherwise, it just becomes abstract numbers.

Finally, it is important that we also seek to validate sentiment and symbolism with data. Modern technology now allows us, through techniques such as natural language processing and machine learning, to determine and quantify the sentiment expressed in communication. We must also use our power of analytical capacity: our human ability to use logic and reason, make connections through logical assessment and mediate our behaviour by the conscious appraisal of events (Slovic, 2007). As the minister further expressed: 'A good leader takes the data, takes the evidence, but also hears the human stories and brings them both together to find how they can best impact.'

## Conclusion

There is no doubt, at least in the minds of most disaster management leaders, that the disaster environment is deteriorating. This is due to a range of factors, including the effects of climate change on natural hazards concurrent with insufficient policy responses to these effects. While our disaster management frameworks and the reviews that help to inform them are attempting to address this deterioration, we risk failing in our duty to lead, serve and protect others if we continue to bias responsibility too far towards citizens within a social contract of shared responsibility for disaster management. Relatedly, we risk falling to the negative effects of being invulnerable if we do not make a greater effort to truly understand how people suffer as well as appreciate what causes them to be vulnerable. We risk losing their trust and confidence and our ability to reduce their suffering and vulnerability.

To ameliorate these risks, we need to put into effect institutional arrangements that rebalance the social contract of shared responsibility. We need to put in place democratic and participative structures that allow for the contestation of ideas about how to provide protection. We need to redistribute the equities of power, wealth and resources made available to governments to regulate, plan, fund, prioritise, facilitate and deliver on community safety needs. We need to address the socio-economic causes of vulnerability, inequity and injustice directly in our resilience policy frameworks. Finally, we need to ensure that economic, scientific and technical knowledge as well as social and cultural knowledge and situational awareness are freely and fully accessible to citizens such that they can make informed choices about the best way to protect themselves.

At the same time, leaders will need to navigate the ethical tensions between principles-based frameworks that seek to impose duties of protection, as well as achieve outcomes that reduce suffering for the greatest number of citizens, and the need to engage relationally with citizens to better understand what is truly important to them. To this end, the character of the leader, premised upon the ethics of trust and compassion and informed by the practical wisdom of their lived experiences, will be fundamental to this success.

There is no easy way out of the dilemma of a deteriorating disaster environment, but there can be a more rewarding experience for all concerned. When, as leaders, we come to realise that our shared humanity is shaped not only by our universal experiences of vulnerability and suffering, but also by our shared capacity to be in relationship with others as we navigate these experiences, we open a unique opportunity. We open the opportunity to place the ethics of trust and compassion in the centre of our leadership ethos, and in so doing, optimise our ability to uphold trust and confidence, and reduce the vulnerability and suffering of those we are called upon to lead, serve and protect.

## Notes

[1] For a more complete picture of national disaster impact over the period visit https://knowledge.aidr.org.au/resources/major-incidents-report/

[2] See https://nema.gov.au/disaster-ready-fund

[3] For example, see *Profiling Australia's Vulnerability: The Interconnected Causes and Cascading Effects of Systemic Disaster Risk* at https://www.aidr.org.au/media/6682/national-resilience-taskforce-profiling-australias-vulnerability.pdf

## References


AIDR (2020) *Major Incidents Report 2019-20*. Australian Institute for Disaster Resilience Melbourne, Victoria, Australia.

AIDR (2021) *Major Incidents Report 2020-21*. Australian Institute for Disaster Resilience Melbourne, Victoria, Australia.

AIDR (2022) *Major Incidents Report 2021-22*. Australian Institute for Disaster Resilience Melbourne, Victoria, Australia.

AIDR (2023) *Major Incidents Report 2022-23*. Australian Institute for Disaster Resilience Melbourne, Victoria, Australia.

Atkinson, C. and Curnin, S. (2020) Sharing responsibility in disaster management policy. *Progress in Disaster Science* 7, 100122.

Barnett, J., Tschakert, P., Head, L. and Adger, W.N. (2016) A science of loss. *Nature Climate Change* 6(11), 976. DOI: 10.1038/nclimate3140.

Beck, U. (1992) *Risk Society: Towards a New Modernity*. Sage Publications, London.

Beck, U. (2009) World risk society and manufactured uncertainties. *Iris* 1(2), 291–299.

Beckett, A.E. (2006) *Citizenship and Vulnerability: Disability and Issues of Social Engagement*. Palgrave Macmillan, Basingstoke, UK.

Bickerstaff, K., Simmons, P. and Pidgeon, N. (2009) Constructing responsibilities for risk: Negotiating citizen – State relationships. *Environment and Planning A* 40(6), 1312–1330. DOI: 10.1068/a39150.

Binskin, M., Bennett, A. and MacIntosh, A. (2020) *Royal Commission into National Natural Disaster Arrangements*. Commonwealth of Australia, Canberra.

Bishara, N.D. and Schipani, C.A. (2009) Strengthening the ties that bind: Preventing corruption in the executive suite. *Journal of Business Ethics* 88(S4), 765–780. DOI: 10.1007/s10551-009-0325-4.

Bruton, S.V. (2004) Teaching the golden rule. *Journal of Business Ethics* 49(2), 179–187. DOI: 10.1023/B:BUSI.0000015783.98570.a3.

Caro, D.H.J. (2015) Code orange: Towards transformational leadership of emergency management systems. *Healthcare Management Forum* 28(5), 195–201. DOI: 10.1177/0840470415588657.

Caro, D.H.J. (2016) The nexus of transformational leadership of emergency services systems. *International Journal of Emergency Services* 5(1), 18–33. DOI: 10.1108/IJES-11-2015-0024.

Carr, D. (2003) Character and moral choice in the cultivation of virtue. *Philosophy (London)* 78(2), 219–232. DOI: 10.1017/S0031819103000251.

Chryssides, G.D. and Kaler, J.H. (1993) *An Introduction to Business Ethics*. Thomson, London.

Cole, L., Eburn, M., Dovers, S. and Gough, M. (2018) Can major post-event inquiries and reviews contribute to lessons management? *The Australian Journal of Emergency Management* 33(2), 34–39.

Cowden, S. and Singh, G. (2017) Community cohesion, communitarianism and neoliberalism. *Critical Social Policy* 37(2), 268–286. DOI: 10.1177/0261018316670252.

Crosweller, M. (2022a) Disaster management and the need for a relational leadership framework founded upon compassion, care, and justice. *Climate Risk Management* 35, 100404.

Crosweller, M. (2022b) Disaster management leadership and the need for virtue, mindfulness, and practical wisdom. *Progress in Disaster Science* 16, 100248. DOI: 10.1016/j.pdisas.2022.100248.

Crosweller, M. and Tschakert, P. (2019) Climate change and disasters: The ethics of leadership. *Wiley Interdisciplinary Reviews: Climate Change* 11(2), 1–18. DOI: 10.1002/wcc.624.

Crosweller, M. and Tschakert, P. (2020) Disaster management leadership and policy making: A critical examination of communitarian and individualistic understandings of resilience and vulnerability. *Climate Policy* 1–19. DOI: 10.1080/14693062.2020.1833825.

Crosweller, M. and Tschakert, P. (2021) Disaster management and the need for a reinstated social contract of shared responsibility. *International Journal of Disaster Risk Reduction* 63, 1–13. DOI: 10.1016/j.ijdrr.2021.102440.

Cunliffe, A.L. and Eriksen, M. (2011) Relational leadership. *Human Relations* 64(11), 1425–1449. DOI: 10.1177/0018726711418388.

D'Agostino, F., Gaus, G. and Thrasher, J. (2019) Contemporary approaches to the social contract. In: *Stanford Encyclopedia of Philosophy*, Fall 2019 edn. Available at: https://plato.stanford.edu/entries/contractarianism-contemporary/ (accessed 11 July 2025)

Dawson, D. (2015) Two forms of virtue ethics: Two sets of virtuous action in the fire service dispute. *Journal of Business Ethics* 128(3), 585–601. DOI: 10.1007/s10551-014-2121-z.

Deloitte (2013) *Building Our Nation's Resilience to Natural Disasters*. Available at: https://www.deloitte.com/content/dam/assets-zone1/au/en/docs/services/economics/deloitte-au-economics-building-resilience-natural-disasters-states-territories-161117.pdf (accessed 11 July 2025).

Deloitte (2016) *The Economic Cost of the Social Impact of Disasters*. Available at: https://alnap.org/documents/9555/report-the-economic-cost-of-the-social-impact-of-natural-disasters.pdf (accessed 11 July 2020).

D'Olimpio, L. (2018) Trust as a virtue in education. *Educational Philosophy and Theory* 50(2), 193–202. DOI: 10.1080/00131857.2016.1194737.

Elliott, E. (2021) Record budget investment to support emergency services. Available at: https://www.budget.nsw.gov.au/sites/default/files/2021-06/ELLIOTT%20-%20Record%20Budget%20investment%20to%20support%20emergency%20services.pdf (accessed 14 July 2025).

Etzioni, A. (1998) *The Essential Communitarian Reader*. Rowman & Littlefield, Lanham, Maryland.

Farmer, P. (2010) *Partner to the Poor: A Paul Farmer Reader*. University of California Press, Berkeley, California.

Faulkner, P. (2018) Finding trust in government. *Journal of Social Philosophy* 49(4), 626–644. DOI: 10.1111/josp.12262.

Fuller, M. and O'Kane, M. (2022) *Flood Inquiry Volume One: Summary Report*. Available at: https://www.nsw.gov.au/sites/default/files/noindex/2022-08/VOLUME_ONE_Summary.pdf (accessed 11 July 2025).

Gergen, K.J. (1999) *An Invitation to Social Construction*. Sage Publications, London.

Gibbs, L., Molyneaux, R., Harms, L., Gallagher, H.C., Block, K. *et al*. (2021) 10 Years Beyond Bushfires. University of Melbourne, Melbourne, Australia. Available at: https://mspgh.unimelb.edu.au/__data/assets/pdf_file/0009/3645090/BB-10-years-report_spread.pdf (accessed 11 July 2025).

Giddens, A. (1990) *The Consequences of Modernity*. Polity Press, Cambridge, UK in association with Blackwell, Oxford, UK.

Giddens, A. (1991) *Modernity and Self-Identity: Self and Society in the Late Modern Age*. Polity Press, Cambridge, UK.

Giddens, A. (1998) *Conversations with Anthony Giddens: Making Sense of Modernity*. Stanford University Press, Stanford, California.

Giddens, A. (1999) Risk and responsibility. *Modern Law Review* 62(1), 1–10. DOI: 10.1111/1468-2230.00188.

Gilson, E. (2011) Vulnerability, ignorance, and oppression. *Hypatia* 26(2), 308–332. DOI: 10.1111/j.1527-2001.2010.01158.x.

Gilson, E. (2014) *The Ethics of Vulnerability: A Feminist Analysis of Social Life and Practice*. Routledge, New York.

Goodin, R.E. (1985a) *Protecting the Vulnerable: A Reanalysis of Our Social Responsibilities*. University of Chicago Press, Chicago.

Goodin, R.E. (1985b) Vulnerabilities and responsibilities: An ethical defense of the welfare state. *The American Political Science Review* 79(3), 775–787. DOI: 10.2307/1956843.

Guyer, P. (2002) Ends of reason and ends of nature: The place of teleology in Kant's ethics. *The Journal of Value Inquiry* 36(2–3), 161. DOI: 10.1023/A:1016140116768.

Handmer, J. and James, P. (2007) Trust us and be scared: The changing nature of contemporary risk. *Global Society* 21(1), 119–130. DOI: 10.1080/13600820601116609.

Hunt, S. and Eburn, M. (2018) How can business share responsibility for disaster resilience? *Australian Journal of Public Administration* 77(3), 482–491. DOI: 10.1111/1467-8500.12320.

IPCC (2014) Summary for policymakers. In: Field, C.B., Barros, V.R., Dokken, D.J., Mach, K.J., Mastrandrea, M.D. *et al.* (eds) *Climate Change 2014: Impacts, Adaptation, and Vulnerability*. Cambridge University Press, Cambridge, UK and New York, pp. 1–32.

Kelman, I. (2020) *Disaster By Choice: How Our Actions Turn Natural Hazards into Catastrophes*. Oxford University Press, Oxford, UK.

Knowles, D. (2001) *Political Philosophy*. Routledge, Abingdon, UK.

Krolikowski, A. (2008) State personhood in ontological security theories of international relations and Chinese nationalism: A sceptical view. *Chinese Journal of International Politics* 2(1), 109–133. DOI: 10.1093/cjip/pon003.

Ladkin, D. (2006) When deontology and utilitarianism aren't enough: How Heidegger's notion of "dwelling" might help organisational leaders resolve ethical issues. *Journal of Business Ethics* 65(1), 87–98. DOI: 10.1007/s10551-006-0019-0.

Li, H.-I. (2017) Rethinking vulnerability in the age of Anthropocene: Toward ecologizing education. *Educational Theory* 67(4), 435. DOI: 10.1111/edth.12264.

Lodge, G.C. (2009) Ideology and national competitiveness. *Journal of Managerial Issues* 21(4), 461–477.

Lukasiewicz, A., Dovers, S. and Eburn, M. (2017) Shared responsibility: The who, what and how. *Environmental Hazards* 16(4), 291–313. DOI: 10.1080/17477891.2017.1298510.

Lupton, D. (2013) *Risk*. Routledge, New York.

MacIntyre, A. (2007/1981) *After Virtue: A Study in Moral Theory*, 3rd edn. University of Notre Dame Press, Notre Dame, Indiana.

McLennan, B.J. and Handmer, J. (2012) Reframing responsibility-sharing for bushfire risk management in Australia after black Saturday. *Environmental Hazards* 11(1), 1–15. DOI: 10.1080/17477891.2011.608835.

McLennan, B.J. and Handmer, J. (2014) *Sharing Responsibility in Australian Disaster Management: Final Report for the Sharing Responsibility Project*. Bushfire Cooperative Research Centre, East Melbourne, Victoria, Australia.

McLeod, J. (2012) Vulnerability and the neo-liberal youth citizen: A view from Australia. *Comparative Education* 48(1), 11–26. DOI: 10.1080/03050068.2011.637760.

Metz, T. and Clark Miller, S. (2016) Relational ethics. In: LaFollette, H. (ed.) *The International Encyclopedia of Ethics*. Wiley. DOI: 10.1002/9781444367072.

Mitzen, J. (2006) Ontological security in world politics: State identity and the security dilemma. *European Journal of International Relations* 12(3), 341–370. DOI: 10.1177/1354066106067346.

Morrell, K. and Dahlmann, F. (2022) Aristotle in the Anthropocene: The comparative benefits of Aristotelian virtue ethics over Utilitarianism and deontology. *The Anthropocene Review* 10(3), 615–635. DOI: 10.1177/20530196221105093.

National Resilience Taskforce (2018) *Profiling Australia's Vulnerability: The Interconnected Causes and Cascading Effects of Systemic Disaster Risk*. Australian Government, Department of Home Affairs, Canberra. Available at: https://www.aidr.org.au/media/6682/national-resilience-taskforce-profiling-australias-vulnerability.pdf (accessed 14 July 2025).

NDRRF (2018) *National Disaster Risk Reduction Framework*. Commonwealth of Australia. Available at: https://www.homeaffairs.gov.au/emergency/files/national-disaster-risk-reduction-framework.pdf (accessed 11 July 2025).

NDRS-NZ (2019) *National Disaster Resilience Strategy*. Ministry of Civil Defence & Emergency Management, Wellington. Available at: https://www.civildefence.govt.nz/assets/Uploads/documents/publications/ndrs/National-Disaster-Resilience-Strategy-10-April-2019.pdf (accessed 11 July 2025).

NSDR-AUS (2011) *National Strategy for Disaster Resilience*. Commonwealth of Australia. Available at: https://www.homeaffairs.gov.au/emergency/files/national-strategy-disaster-resilience.pdf (accessed 11 July 2025).

O'Connell, D., Wise, R.M., Doerr, V., Grigg, V., Williams, N. *et al.* (2018) *Approach, Methods and Results for Co-producing a Systems Understanding of Disaster: Technical Report Supporting the Development of the Australian Vulnerability Profile*. CSIRO, Canberra. Available at: https://publications.csiro.au/rpr/pub?pid=csiro:EP187363 (accessed 14 July 2025).

Overman Smith, E. and Thompson, I. (2002) Feminist theory in tactical communication: Making knowledge claims visible. *Journal of Business and Technical Communication* 16(4), 441.

*Oxford English Dictionary* (2017) Trust. Available at: https://www.oed.com/ (accessed 14 July 2025).

Poppo, L., Zhou, K.Z. and Li, J.J. (2016) When can you trust "trust"? Calculative trust, relational trust, and supplier performance. *Strategic Management Journal* 37(4), 724–741. DOI: 10.1002/smj.2374.

Quinn, M. (2017) Place leadership and the social contract: Re-examining local leadership in the East Midlands. *Local Economy* 32(4), 281–296. DOI: 10.1177/0269094217707279.

Ridley-Duff, R. (2007) Communitarian perspectives on social enterprise. *Corporate Governance: An International Review* 15(2), 382–392. DOI: 10.1111/j.1467-8683.2007.00568.x.

Rogers, B. and Frederick, C. (2000) Beyond "Identity". *Theory and Society* 29(1), 1–47. DOI: 10.1023/A:1007068714468.

Rousseau, D.M., Sitkin, S.B., Burt, R.S. and Farrell Cameraer, C. (1998) Not so different after all: A cross-discipline view of trust. *Academy of Management Review* 23(3), 393–404.

Singh-Peterson, L., Salmon, P., Baldwin, C. and Goode, N. (2015) Deconstructing the concept of shared responsibility for disaster resilience: A Sunshine Coast case study, Australia. *Natural Hazards* 79(2), 755–774. DOI: 10.1007/s11069-015-1871-y.

Slovic, P. (2007) "If I look at the mass I will never act". *Judgment and Decision Making* 2, 79–95.

Sobhy, H. (2021) The lived social contract in schools: From protection to the production of hegemony. *World Development* 137, 104986. DOI: 10.1016/j.worlddev.2020.104986.

Szerszynski, B. (1999) Risk and trust: The performative dimension. *Environmental Values* 8(2), 239–252. DOI: 10.3197/096327199129341815.

Terblanche, N.H.D. and Heyns, M. (2020) The impact of coachee personality traits, propensity to trust and perceived trustworthiness of a coach, on a coachee's trust behaviour in a coaching relationship. *SA Journal of Industrial Psychology* 46(1), 1–11. DOI: 10.4102/sajip.v46i0.1707.

Tschakert, P., Barnett, J., Ellis, N.R., Lawrence, C., Tuana, N. *et al.* (2017) Climate change and loss, as if people mattered: Values, places, and experiences. *Wiley Interdisciplinary Reviews: Climate Change* 8(5), 476–495. DOI: 10.1002/wcc.476.

UNISDR (2015) *Sendai Framework for Disaster Risk Reduction 2015-2030*. The United Nations Office for Disaster Risk Reduction, Geneva, Switzerland. Available at: https://www.preventionweb.net/files/43291_sendaiframeworkfordrren.pdf (accessed 14 July 2025).

Velasquez, M.G. (2001) *Business Ethics: Concepts and Cases*. Prentice Hall, London.

Wachinger, G., Renn, O., Begg, C. and Kuhlicke, C. (2013) The risk perception paradox—implications for governance and communication of natural hazards. *Risk Analysis* 33(6), 1049–1065. DOI: 10.1111/j.1539-6924.2012.01942.x.

White, M.J. (2012) *Political Philosophy: A Historical Introduction*, 2nd edn. Oxford University Press, New York.

Wisner, B., Gaillard, J. and Kelman, I. (2012) *The Routledge Handbook of Hazards and Disaster Risk Reduction*. Routledge, Abingdon, UK.

# 4 Addressing the Wicked Problem of Systemic Disaster Risk in the Anthropocene

**John Fien* and Esther Charlesworth**

*School of Architecture and Urban Design, RMIT University, Natural Hazards Research Australia, Melbourne, Australia*

**Abstract**

The Anthropocene is characterised by the interconnected problems of economic disruption, urban decay, inequality, political inertia, land degradation, and climate change. Together, these constitute a polycrisis, which is multiplying the impacts of hazard risks and increasing the potential for disasters. However, many communities fail to recognise these interconnections and are largely unprepared when disasters strike, or do not appreciate that the heaviest burdens of disasters mostly fall on poor and vulnerable people who have limited resources to cope. Too often, engineering works, such as stream straightening, levee banks, and sea walls against floods and sea-level rise, are seen as the solution. While these can mitigate immediate impacts, they fail to address the root causes of a disaster, which reside not in the environmental hazard *per se* but in the socio-economic and political drivers of vulnerability. Thus, it is often said that disasters are a function of 'systemic risks' that create wicked problems. This chapter begins by outlining the complex nature of systems and systemic risks in the Anthropocene and how these are leading to increasing compound disasters and cascading impacts that reverberate through systems. This means that disaster mitigation and recovery need to be treated as wicked problems. The chapter explains the characteristics that make disaster risk management wicked and the challenges of seeking resolutions to them. Nevertheless, a range of strategies for addressing these challenges is emerging, including appreciative enquiry, community-based adaptation, strategic foresight, and systemic design. The chapter concludes by outlining the principles shared by these strategies and illustrates these with the example of the steps involved in systemic design as a meta-approach to addressing the wicked problem of disaster risk in the Anthropocene.

## Introduction

The old maxim that 'It never rains but it pours' is a condition of life in the polycrisis. So, it is not surprising that it is also true of the occurrence and impacts of environmental hazards where the acute polycrisis conditions of interconnected problems of economic disruption, urban decay, inequality, political inertia, land and biodiversity degradation and climate change are multiplying the impacts of hazard risks. In fact, environmental hazards are a polycrisis in miniature.

A single disaster – a cyclone, a flood or an earthquake, for example – can leave devastated homes and livelihoods, severely disrupted critical infrastructure and services and a significant number of serious injuries and fatalities, and can be the cause of overwhelming grief and trauma. The reality and severity of disaster impacts are generally starker than this, however, due to the systemic nature of risk through

*Corresponding author: john.fien@rmit.edu.au


DOI: 10.1079/9781800624153.0004

which two or more environmental hazards often occur in close succession. For example, the Los Angeles wildfires in January 2025 were a series of discrete fires that broke out in multiple suburbs over a number of weeks (Pequeño and Ray, 2025) while the heavy rain and destructive winds brought by a tropical cyclone – and six of these, for example, hit Metro Manila in The Philippines in less than a month in late 2024 (OCHA, 2024) – will very often be followed by floods, landslides and coastal storm surges. In the same vein, the damage of the 'Black Summer' of bushfires in Eastern Australia in 2019–2020 was soon followed by cyclones – with rescue and relief workers having to respond to injuries and the loss of life, damage to infrastructure and economic upheaval, all under the restrictions of the COVID-19 pandemic (Gissing *et al.*, 2022).

In addition, endemic stressors, such as ill-advised planning decisions and inadequate building codes, together with insecure employment and health inequalities, have made many communities particularly vulnerable to the effects of hazards and their impacts, especially as they cascade across social, economic and environmental systems. The resulting prolonged period of disaster recovery and increasing disaster risk also means that many people remain unduly vulnerable to further disasters while they are still in a vulnerable recovery phase from an earlier one.

These hazard-based risks are growing in frequency, impact and scale and have intensified so much that, today, even exposure to non-extreme hazards can produce extreme impacts. The disaster sociologist Susan Cutter (2021, p. 822) argues that 'extreme events' are being turned into 'annual, seasonal, monthly, or everyday occurrences' while

> Nuisance or sunny-day flooding in coastal areas used to be unusual with the annual king tide but has now transformed into a routine occurrence, producing chronic flooding conditions that disrupt daily activities, affecting public and private infrastructural systems such as roads and sewers, damage property, and most significantly, threaten public safety. (Cutter, 2021, p. 822)

Thus, the Rockefeller Foundation (2024) argues that many communities are 'largely unprepared to respond, withstand, and rebound' when the disasters strike – and that the greatest burden of disasters mostly falls on poor and vulnerable people who have limited resources to cope and who take longer to recover from them, thus increasing inequality. Too often, engineering works, such as stream straightening, levee banks and sea walls against floods, are seen as the solution, but, while these can mitigate immediate impacts, they fail to address the root causes of a disaster, which reside not in the environmental hazard *per se* but in the socio-economic and political drivers of vulnerability.

Four concepts are important in understanding ways of addressing such complex disaster risks effectively. These are: the Anthropocene, complex systems, wicked problems, and systemic design. These four ideas guide the structure of this chapter and, together, are used to develop a meta-approach to addressing the wicked problem of disaster risk in the Anthropocene.

## The Polycrisis and the Anthropocene

The 'polycrisis' is said to be a buzzword whenever business, government and community leaders meet to discuss the many problems in the world today (Drezner, 2023; Eveleigh, 2023). For example, the theme of the 2023 World Economic Forum in Davos was 'Cooperation in a Fragmented World', which sought to address the 'sheer number of ongoing crises' that have placed Earth and her people 'at a critical inflection point' (WEF, 2023a). Documents presented at the Forum argued that:

> The twin triggers of the COVID-19 pandemic and the war in Ukraine rattled an already brittle global system. Economic growth in the world's largest economies is stalling, while navigating headwinds from rising food and energy prices . . . This is occurring alongside heightened geo-economic fragmentation, financial sector vulnerabilities, including stretched asset prices and high debt levels, and a climate crisis spiralling out of control, which could magnify any growth slowdown, particularly in emerging markets. Unless these systemic and interconnected risks are addressed, the promise of a 'decade of action' may become a decade of uncertainty and fragility. (WEF, 2023b)

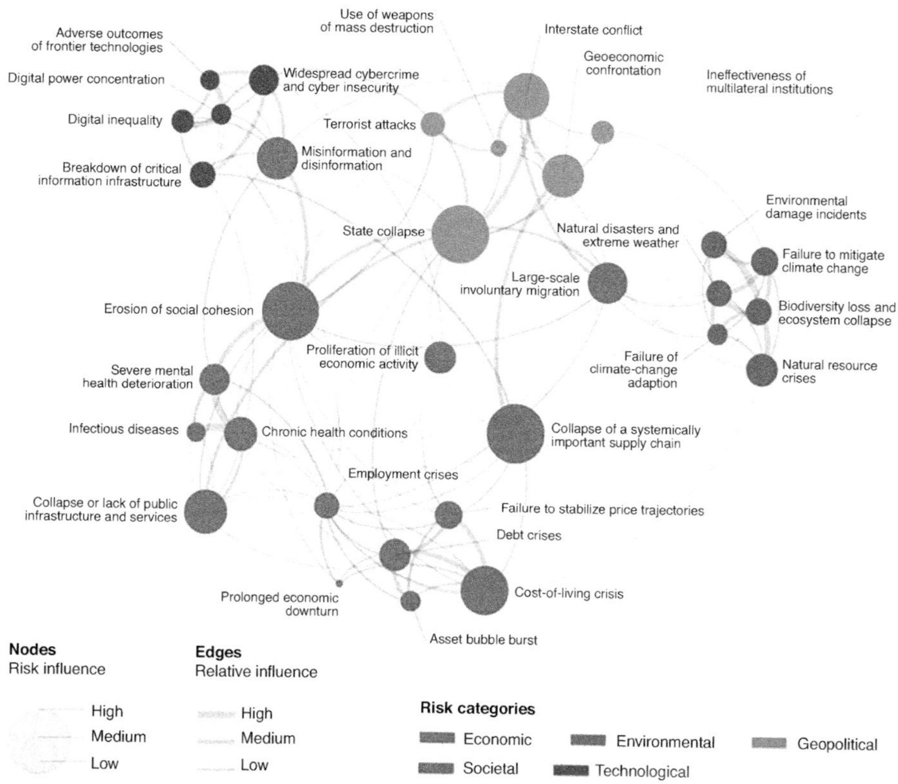


**Fig. 4.1.** The multiple and intersecting crises that collectively constitute the polycrisis. From WEF, 2023c.

This and related meetings of World Economic Forum were a response to the findings of its annual *Global Risk Report* that identified the multiple and interconnected crises facing the world. Looking at the complex interactions depicted in an infographic from the report (see Fig. 4.1), it is not surprising that 'polycrisis' is such a much-used term today.

But what is a (or *the*) polycrisis? The term was originally coined by philosopher Edgar Morin in *Terre Patrie*, which was written with Anne Kern and translated into English as *Homeland Earth: A Manifesto for the New Millennium – Advances in Systems Theory, Complexity, and the Human Sciences* in 1999 (Morin and Kern, 1999). They used the term to describe the situation that results when multiple, simultaneous crises interact in complex systems and become entangled in ways that compound and reshape one another. Being more than the sum of their parts, systemic crises can produce impacts greater and more dangerous than the sum of the risks the individual crises would produce if their host systems were not so deeply interconnected (Lawrence *et al.*, 2022).

These interconnected and compounding crises make the polycris the defining characteristic of the Anthropocene, the geological epoch into which Earth is said to have now entered. The term Anthropocene indicates that the human species is now the primary driver of change in Earth's biophysical systems. This is unlike the Holocene epoch (the 11,600 or so years since the last ice age) and the preceding millions of years of different geological epochs and eras. The Holocene has been the only period in Earth's history when the environmental state of the planet was able to support very large numbers of people to live in settled societies and thrive (Meyer and Newman, 2020). The Anthropocene is thus a volatile, uncertain, complex and ambiguous context for humanity and its supporting social, built and natural systems (Miyagawa and Ohta, 2022). However, it is unknown whether the polycrisis will enable this condition to continue or whether the risks and vulnerabilities it brings to the Anthropocene will breach planetary boundaries and contribute to systems disruption and decline (Lenton *et al.*, 2019; Lawrence, 2024). The uncertainty about systemic risks is what, in the end, differentiates complex systems from simple ones (Lennox, 2024).

The scientific consensus seems to be pessimistic on the issue of life continuing as it was in the Holocene into the Anthropocene. For example, a recent article in *Nature* describes the current situation as one of elevated risk, in which:

> . . . the planet is now hotter, more contaminated and biologically more degraded. These negative trends are set to intensify and extend further outside the Holocene envelope. Some of the changes involved are long lasting (such as climate change) and some are irreversible (such as extinctions). They are already exerting pressure on political institutions, legal frameworks and economic relations, all of which are meant to protect human communities and give them meaning. (Zalasiewicz *et al.*, 2024, pp. 981–982)

It can also be noted that our 'political institutions, legal frameworks and economic relations' are additionally, and at the same time, enabling development decisions that disrupt the functioning of biophysical systems, such as water, carbon, energy and ecological cycles, and bring us to the systemic 'tipping points' that lead to eventual collapse. Scheffran (2016) presented the systemic drivers of such vulnerabilities and risks diagrammatically. Using the arrows in Fig. 4.2 to depict the interactions within and across systems, he explained that changes in any of the systems and their subsystems can undermine the functionality and stability of all the others. For example, a single, localised weather event, let alone the climate system itself, can cause major impacts in all systems if critical tipping points, such as limits to temperature increases, for example, are reached. As the resulting impacts cascade and multiply across the systems, the chains of events can lead to global and lasting transformations that can only be anticipated as possibilities rather than be predicted scientifically. Breaching tipping points risks the possibility of planetary destabilisation, given the significant uncertainties that result from the domino effect of impacts in complex systems (Ripple *et al.*, 2023).

The key challenge facing those concerned about such conditions leading to systems collapse relates to how changes can me made to 'political institutions, legal frameworks and economic relations' in the Anthropocene to protect human communities and the biophysical systems upon which social and economic development depends.

These conditions are not only the results of interactions within, between and across coupled biophysical and social systems. They are also drivers of current and future risks. Risks from natural hazards are increasing considerably under such conditions. For example, disasters related to hazards killed almost 100,000 people and displaced over 90 million more worldwide in 2023. The economic losses from these exceeded A$300 billion (CRED, 2025). Under such conditions, stressors and shocks intertwine to be so inextricably bound together to be considered 'systemic risks'. Systemic risks particularly predominate in areas with high levels of vulnerability. Together, these aspects of systemic risk make multiple or compound disasters very likely to occur in close spatial and/or temporal proximity. They also generate impacts that are likely to cascade and intensify. As a result, these interlocking risks can be said to be 'accelerating, amplifying, and increasingly synchronizing with one another in ways we do not understand but must urgently investigate if we are to effectively address the world's intersecting problems' (Lawrence, 2024, p. 1).

## Complex Systems

From the preceding section, it is clear that we do not live in simple systems, especially when the challenges of the Anthropocene are factored in. Fig. 4.1 identifies some of the distinguishing features of simple and complex systems, clearly illustrating that disaster risks are deeply embedded within the complex interactions and interdependence of natural and human systems

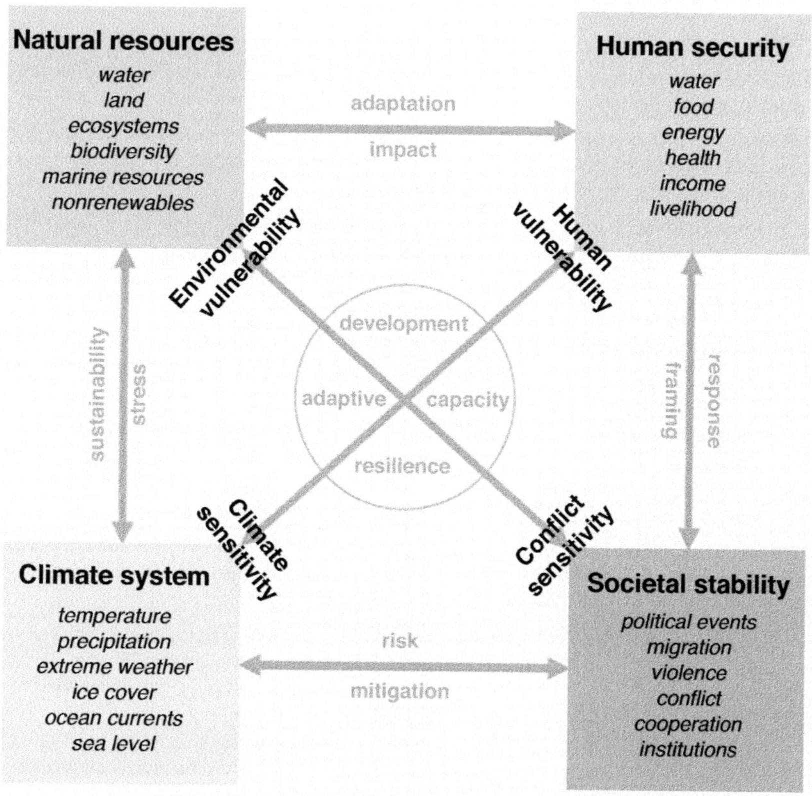


**Fig. 4.2.** The nexus of systemic risks created by interactions and feedback in coupled biophysical and social systems. From Scheffran, 2016 p. 310. Reproduced with permission from Springer Nature.

and their respective subsystems. Comprised of all the interacting systems and subsystems that support life in all its forms in ecosystems and biomes, natural systems include geomorphological, atmospheric, ecological, fresh water and marine processes. Human systems are comprised of all the relationships between individuals and groups in a society, such as purely social relationships as well as the economic, cultural, infrastructural and political ones that sustain human life in cities and give it structure and meaning. The interplay of interconnected natural and social systems has often led human society and its environments to be described as 'systems of systems of systems', thus making them among the most complex systems on Earth (Table 4.1; de Roo *et al.*, 2020).

This insight is not particularly new, especially to those who have been working in these fields for some time (Burger *et al.*, 2021). However, despite the increasing recognition of the complexity and the coupled nature of risks in the Anthropocene, very often urban planners, for example, 'do not see disasters and disaster prevention as being within their remit' while 'few national and international disaster agencies have worked with urban governments and community organizations to identify and act on the ... processes that cause the accumulation of disaster risk in and around urban areas' (Bull-Kamanga *et al.*, 2003, p. 193).

As a result, many working in the field of resilience to hazards and risks continue to apply simple, independent, 'one-off' solutions based upon engineering rather than socio-ecological and transformative views of resilience (see Table 4.2). Indeed, from their analyses of engineering resilience projects to mitigate riverine

**Table 4.1.** Characteristics of simple and complex systems. Adapted from Riley *et al.*, 2017.

| Simple systems | Complex systems |
|---|---|
| Homogeneous: Identical/indistinguishable structural elements in the system. | Heterogeneous: A large number of structural variations in system elements. |
| Independent elements and subsystems: These interact but do not influence or control other parts of the system. | Interdependent elements and subsystems: These do not just interact but are interconnected or interwoven. |
| No feedback loops: Little feedback to elements or subsystems from the impacts of their interactions. | Feedback loops: Multiple feedback loops ensure elements and subsystems are in constant flux from a chain of causes and impacts from the interconnections. |
| Static: Little changes in the system over time. | Dynamic: Significant changes over time with the past impacting on the present and the present on the future. |
| Linear relationships: Interactions produce one-for-one direct effects. | Non-linear relationships: Interactions between multiple interconnected elements and subsystems produce compound and cascading impacts not proportional to the causes. |
| Deterministic and predictable: The same result always occurs for a given set of interactions. | Stochastic: Non-linear relationships produce significant randomness leading to uncertain and unpredictable outcomes. |
| No adaptation, self-organisation or emergence: The elements and subsystems and their interactions remain static. | Adaptation and self-organisation: The dynamic and stochastic changes within complex systems facilitate a restructuring (or self-organisation) of the elements, subsystems, interactions and/or interconnections within the system as they adapt to new conditions and create new structures, learn and/or diversify. |
| | Emergence: New patterns or processes can emerge from adaptations and self-organisation within a system, and which can create new systemic risks and/or new systemic capacities for systemic resilience. |

and coastal flooding in Bangkok, Jakarta, Manila and Ho Chi Minh City, Yarina (2018) and Goh (2021) conclude, for example – and not metaphorically – that 'Your sea wall won't save you'. This is because these and related projects tend to operate from the perspective of simple systems, such as building a sea wall against land encroachment from rising sea levels, and to continue the sea wall example, without recognising historic inequalities, the loss of beach access to their boats required by fisherfolk, land subsidence from groundwater removals (meaning the land is falling beneath sea level, not just the seas rising) and related socio-economic bases of social injustice (also see Nunn *et al.*, 2021). In all the cases that Yarina and Goh analysed, they found that the rivers and seas found ways around the sea walls and levee banks and the impacted communities had to adapt to the new patterns of vulnerability, often with women leading ways of responding to make their homes more disaster resilient.

These various ways through which systems respond to risks and disasters make them a particular type of complex system – a complex adaptive system – in which they operate as a dynamic entity 'capable of evolving and adapting to the environment' (Alaimo, 2021, p. 2). A particular characteristic of complex adaptive systems is this capacity to self-organise and – continuously – reorganise both their diverse components and the patterns and impacts of their ongoing interactions, especially as disruptions occur (Coetzee *et al.*, 2016). New systemic

**Table 4.2.** Characteristics of engineering, socio-ecological and transformative resilience. Adapted from Cretney, 2014; Cinner and Barnes, 2019; Glaeser, 2022; Hassink and Chu, 2025.

| Forms of resilience | Focus, purpose and mechanisms |
|---|---|
| Engineering | Engineering resilience is sought through material interventions that seek to make a system 'fail-safe'. It does this by increasing the resistance of a system to disturbance and is measured by the speed by which the system returns to stable state from a stressed state. The faster the system bounces back, the more resilient it is, with an emphasis is on robustness and efficiency in returning to a pre-disturbance state. |
| Socio-ecological | Social-ecological resilience is the capacity of coupled human and natural systems to absorb shocks and respond to the consequent changes by adapting (or reorganising) system elements, subsystems and their interconnections in order to retain essential functions without fundamentally altering their core identity. |
| Transformative | Transformative resilience emphasises the capacity of a system to transform in response to crises or opportunities, leading to a new system patterns and processes that better address underlying vulnerabilities and promote long-term sustainability. Thus, it goes beyond adaptation and involves fundamentally changing the system to create new pathways and structures that are safer, and more sustainable and equitable. |

relationships, structures and processes result or emerge from these emergent dynamics and then feed back into the 'system of systems' to influence future interactions – and so on. As Folke (2016) explains, '[t]he properties of complex adaptive systems change because of the interplay between the adaptive responses of the parts [of the system] and the emergent properties of the whole that then feed back to the parts' (pp. 5–6). The adaptations that result co-evolve with the systemic processes of interacting feedback loops, time lag effects, thresholds and tipping points, making the outcomes of adaptation quite unpredictable (Liu *et al.*, 2021). Human interventions that seek to strengthen system resilience by guiding or managing the course of self-organisation and adaptation to natural hazard risks need to be mindful of these systemic complexities which make them 'wicked problems'.

## Wicked Problems

The difficulty in addressing resilience challenges does not lie in the magnitude of the damage of any particular disaster or in the strategies for managing their impacts. Rather, the 'wickedness' of the challenge resides in managing the risks posed by the hazards in complex social-ecological systems, where for example, increasingly multiple or compound disaster events simultaneously impact the elements, subsystems and their interconnections in a region. Managing disaster risks and vulnerabilities is a wicked problem, indeed a 'super wicked' problem (Levin *et al.*, 2012).

The concept of wicked problems arose in the work of two urban planners, Rittel and Webber (1973), who recognized that rational analyses followed by logically planned interventions were not amenable to the complex socio-ecological problems that characterised their work in cities. They called these 'wicked problems' and identified a range of ways in which they contrasted with the 'tame' sort of problems found in simple systems. Other terms used to describe the wickedness of such problems include 'intractable controversies' and 'complex', 'unstructured', and 'incorrigible' problems. Wicked problems are found in, and indeed, caused by, the complexity of coupled human and natural systems (often called CHANS – see Liu *et al.*, 2021) and thus have no clear definition or obvious, expert solutions, as in simple systems. Similarly, the complexity of systems makes it extremely difficult to know in advance if any proposed solution will reduce risks and improve matters or, conversely, make matters worse. Overall, and summarising Peters (2017) and Head (2022), wicked problems have many of the following traits:

- Every wicked problem is essentially unique and often constantly changing.
- Wicked problems are difficult to define and are described differently by various stakeholders.
- There are multiple explanations for the wicked problem, with stakeholders often disagreeing strongly about causes and solutions.
- Every wicked problem may be a symptom of another problem.
- Wicked problems have no clear solution or set of possible solutions.
- A focus on short-term solutions may be addressing the symptoms of wicked problems instead of the underlying causes.
- The knowledge base required for effective resolution may be weak, fragmented or contested.
- Some solutions may depend on achieving major shifts in attitudes and behaviours; however, such shifts may be too difficult owing to:
    - a lack of incentives or points of leverage for change;
    - no immediate or ultimate test for solutions.
- Solutions to wicked problems are not true or false, only more or less appropriate or inappropriate.

Super wicked problems have the additional characteristics of:

- Extreme urgency, with a sense of time running out.
- Short-term solutions are given priority given both this sense of urgency and a prevailing decision making socio-economic context that discounts the future.
- Short-term solutions do not solve the problem or even reach their potential and may actually exacerbate problems.
- A lack of a central authority to make decisions and a lack of experience in distributed leadership and collaborative governance.

Given such a list of characteristics, the wicked problems we find in the Anthropocene have been described as growing 'to levels of existential risk' with 'human limits to cope . . . reached or breached' (Jones, 2014, p. 92). Fortunately, in the 50 years since Rittel and Webber raised awareness of the nature of wicked problems, numerous programmes across many disciplines have sought to find ways of working with and around them, with the consensus conclusion being that transdisciplinary approaches are needed as no one discipline is adequate for the challenge of understanding and acting, not just on natural and human systems, but on the tightly coupled and cascading impacts of the interactions between them (Hou *et al.*, 2022).

Others have sought to list and categorise the types of approaches possible. For example, Head (2022) has identified seven: avoidance and denial, trusting the problem will go away; coercive control by an authoritative leader; engaging expert problem-solvers; micro-managing the problem by breaking it down into smaller, 'tamer' parts; making incremental adjustments; developing coping strategies; and wider stakeholder collaboration. Roberts (2000) and Rayner (2006) have classified these and similar approaches into three categories, competitive, authoritative and collaborative, while Matyas and Pelling (2014) categorise the goals of different approaches to wicked problems as being resistance, incremental adjustment or transformation. However, it is only the third of Roberts's and Rayner's categories (or Head's 'wider stakeholder collaboration') that has widespread endorsement in the literature. As Brink and Wamsler (2018) argue, building resilience to the interlocking risks posed by wicked urban, biodiversity, climate, disaster and similar problems requires collaborative governance and the related practices of public participation, public–private partnerships, cross-sector partnerships, cooperative or 'hybrid' governance and the co-creation and co-production of knowledge, understanding and potential solutions.

The 2024 *Navigating New Horizons* report, jointly produced by United Nations Environment Programme (UNEP) and the International Science Council (ISC), synthesised these many approaches into three broad principles, with the priority actions recommended for each one listed in Table 4.3.

**Table 4.3.** Strategies and priority actions for integrating and promoting planetary health and human wellbeing. Adapted from UNEP–ISC, 2024 pp. 63–68.

| Strategies | Actions |
|---|---|
| Develop a new social contract based upon shared values. | • Embrace and engage the views of a diverse group of stakeholders.<br>• Identify marginalised social groups and provide them with the tools and opportunities to have a stronger voice.<br>• Collaborate in developing a new concept and framework for prosperity.<br>• Shape all activities to increase levels of social capital. |
| Develop agile and adaptive governance arrangements and decision making processes. | • Facilitate communities to engage and experiment with and learn by and from co-creating alternative visions and possible and preferred visions and strategies.<br>• Collaboratively develop a structure for distributed leadership and appropriately shared decision making.<br>• Frame all interventions through an agreed theory of change and a clear logic model of inputs, outcomes and impacts, supported by a strong set of social and environmental indicators and targets for monitoring and evaluation. |
| Guide governance decisions through evidence-based data and knowledge. | • Develop platforms for combing data, information and knowledge, and making it accessible to all stakeholders (safeguarding individual privacy).<br>• Develop strong plans for monitoring and evaluating (M&E) all systems interventions and integrate them into project management processes.<br>• Conduct regular opportunities to review M&E findings and integrate discussions into 'lessons learnt' workshops to understand progressive outcomes and impacts from the intervention.<br>• Develop effective communication strategies for taking the results to all stakeholders.<br>• Build capacity for analysing lessons learnt and making decisions for revising the intervention. |

1. Develop a new social contract based upon shared values.
2. Develop agile and adaptive governance arrangements and decision making processes.
3. Guide governance decisions through evidence-based data and knowledge.

There are several examples of success in redressing the wicked problem of reducing disaster risks and building resilience to them by following such principles. Several successful case studies from Sweden (Brink and Wamsler, 2018) and Australia (McNaught *et al.*, 2024) have verified this. However, despite these cases, practical guidance on actually what to do and how to facilitate collaborative governance or to plan and implement action plans, strategies and techniques to develop values consensus is under-represented (to say the least) in much professional practice, in the training of policy and of front-line staff to use such strategies and techniques and in the academic literature. Reflecting the need for systemic methods to address systemic risks, Lee (1993) and Jones (2014) have identified six characteristics of what such operational strategies need to entail if socio-ecological and transformative approaches to resilience are the goal:

- *human-centredness* to facilitate understanding of human concerns, plans and the benefits they seek;
- *engaging diverse stakeholders* to facilitate a personal stake in the outcome of the strategy(ies) being planned;

- *structured processes of dialogue* to enable authentic interaction in collaborative decision making and enable participants to achieve a collective systems view and agreed, if diverse, conclusions;
- *iterative inquiry* that enables individual and social learning from participation, the integration of new thinking and the development of adaptive capacities;
- *integrated and coordinated actions* within natural, social and economic systems and across their drivers and interconnections;
- *treating strategies and plans as experiments* to be closely monitored over time, with new interventions developed and tested.

## Systemic Design

*Knowing how*, as well as *knowing about and why* (Ryle, 2009/1949), is essential for going beyond the incremental changes that address the manifestation or symptoms of wicked problem rather than their root, systemic causes. The purpose of this final section of the chapter is to identify examples of practical strategies based upon these operational characteristics of addressing the wicked problems of disaster risk in the Anthropocene.

Among the participatory strategies and tools that have been applied in achieving this goal are the following:

- participatory learning and action (Kindon *et al.*, 2007; Napier and Simister, 2017);
- action planning for cities (Hamdi and Goethert, 1997; MCR2030, n.d.);
- appreciative enquiry (Whitney and Trosten-Bloom, 2010);
- community-based adaptation / resilience planning (AIDR, 2020; Nath, 2024);
- strategic foresight (Hurtado and DeAngelis, 2022; OECD, 2025);
- design thinking (Lee, 2020; Pedgley and Şener, 2024);
- scenario planning and monitoring (Riddell *et al.*, 2019; Sarkar-Swaisgood and Dewi, 2020);
- participatory landscape urbanism (Elgizawy and Elewa, 2022; Yang *et al.*, 2024).

All these approaches have four features in common. First, they are collaborative, working from the bottom up with 'experts' consciously taking on the role of resource persons, facilitators and guides. Second, they involve a detailed analysis and mapping of the systems and subsystems, and the interconnections within and across them, which frame the root causes, not symptoms, for action. Third, they encourage creative and divergent thinking to address as many parts of the systems as required. Fourth, they recognise the dynamic and emergency nature of systems and, thus, are never seen as a 'final solution'. Rather, as Lee (1993) argued, they need to be closely monitored and then revised, abandoned or consolidated, depending on the desirability, or otherwise, of impacts.

These four features are traits of an emerging strategy known as systemic design, which is an iterative integration of systems thinking and design thinking. This requires action on wicked problems to be framed within a problem-solving process within a systemic thinking perspective. This means looking at the system-as-a-whole, analysing components and subsystems and focusing particularly on the nature and impacts of all interlinkages. In practical terms, this involves zooming in to individual components of the system, examining their connections with others and then zooming back out again to the whole. From this analysis, a range of potential interventions and scenarios can be brainstormed, discussed, evaluated for their desirability and their feasibility and then action plans developed for implementation and monitoring. These processes thus integrate systems thinking into designer-ly 'problem-solving methods and visualization practices and creates innovative reconfigurations for complex services and systems' (Nohra *et al.*, 2020, p. 5).

Systemic design can take many forms, but a key guide is the *Systemic Design Toolkit* (Van Ael *et al.*, 2021), which has seven main steps, as in Fig. 4.3. The iterations between systems thinking to frame a problem and design thinking to create the potential interventions are briefly outlined below.

1. Frame the system: mapping the systems and subsystems to generate a shared understanding of the context of the problem or challenge, to understand current practices,

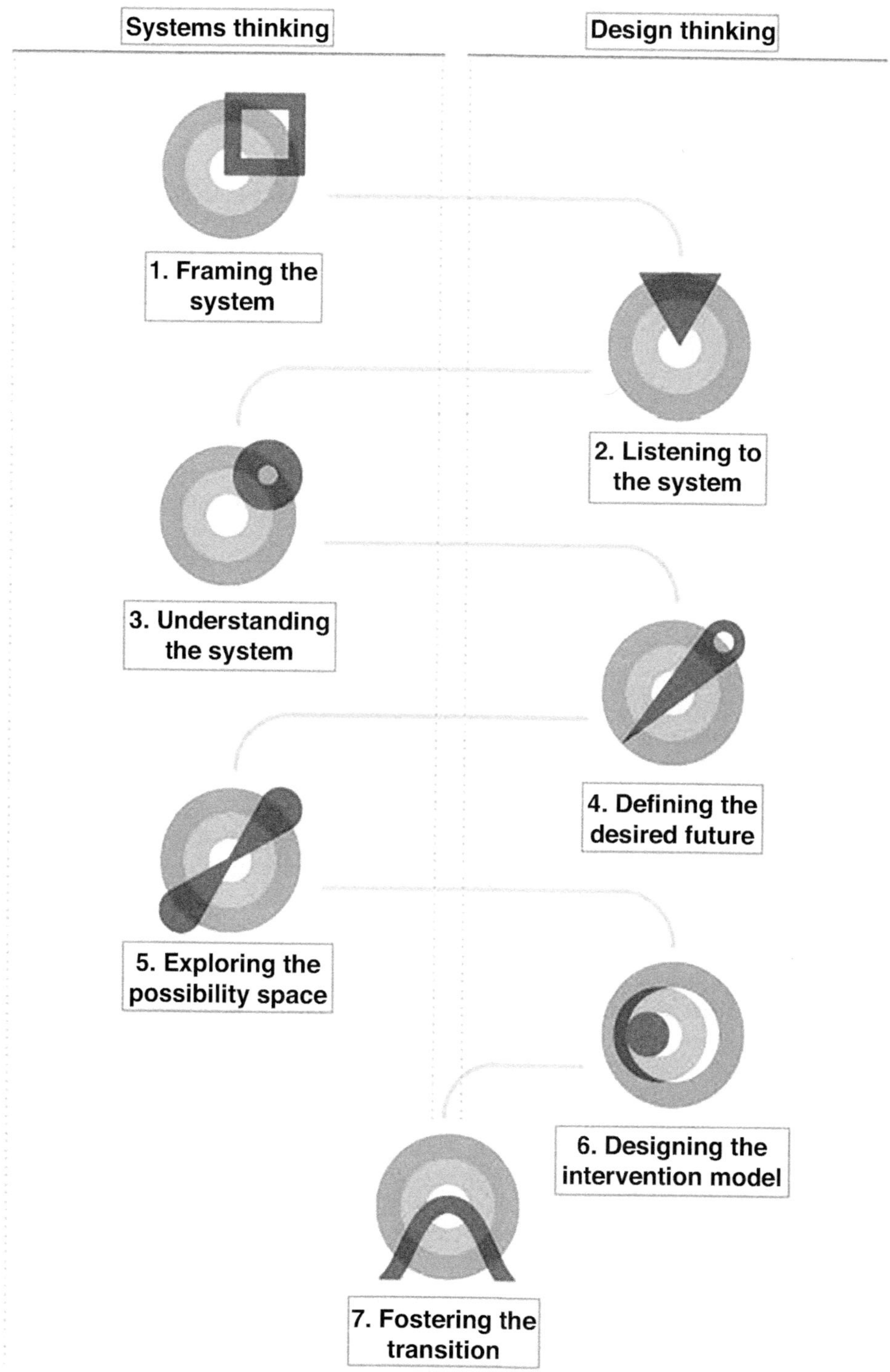


**Fig. 4.3.** Seven steps in systemic design. From Van Ael *et al*., 2021.

trends and innovative initiatives and to identify the actors (stakeholders) to involve. Note: From more-than-human perspectives, features in the natural and built environment can be considered as 'actors' (see Edwards *et al.*, 2023).

2. Listen to the system: with empathy, humility and an appreciation of cross-cultural and cross-species perspectives, analyse the interactions between the actors, by identifying hidden relationships, particularly nature–human interdependencies and the impacts of inequality on the systems and subsystems.
3. Understand the system: develop a shared understanding about forces and interdependencies in the system to discover the leverage points (Fischer and Riechers, 2019) at and through which transformative change may be possible.
4. Define desired future/s: co-design ideal, desired futures by imagining how we want to improve the future context of individuals, society and natural systems. What is the future we are imagining? What do we want to change and how?
5. Explore the possibility space: a brainstorming activity to give sense to the whole process needed to be explored to identify different types of possible interventions by making sure they are covering the big picture of the system that emerged in the initial steps.
6. Design the intervention strategy: assess the desirability and then the feasibility of the potential interventions, making sure to investigate how various desired interventions may connect and possibly reinforce each other to develop an effective action plan for change.
7. Implement and monitor the transition plan: take collaborative steps in the action plan towards the desired goals so that the implementation of the interventions occurs in such a way that transformation can happen step by step – all the while carefully monitoring for successes and any questions, issues or problems that may arise so that the process is cyclical and feeds back into previous leverage points and/or steps.

Following such steps enables communities to co-evolve with natural systems and subsystems to address the disaster risks that emerge from the interconnections with which they live in complex adaptive systems. Indeed, Nelson (2022) argues that 'Systemic design is the preeminent choice for securing desired outcomes in environments of uncertainty and complexity because systemic designing necessarily takes broader stances and deeper approaches . . . than has been the norm'.

## Conclusion

The brief survey in this chapter has introduced some of the ways in which the accelerated and cascading impacts of disaster risks are a dominant aspect of life in the Anthropocene. The tight coupling of biophysical and human processes in the complex adaptive systems that underpin life in the Anthropocene has created a polycrisis of interlocked wicked problems. The chapter has also identified that wicked problems are slippery and have no easy solutions. They cannot be solved through the same type of simple systems resilience that caused them and results in solutions to today's problems being the cause of further wicked problems tomorrow.

Rather, subtle and sophisticated approaches to working with wicked problems of disaster risk management are needed. Research into the wickedness (i.e. the devastating impacts) of disaster problems has taught us that the injuries, loss of life, infrastructural damage and displacement they bring cannot be mitigated by managing the emergency when it happens. National and international strategies have shifted away from trying to manage the disaster itself to *managing the risk* of any disaster occurring, mitigating any destructive impacts and building disaster resilience through recovery programmes that develop the adaptive capacities of infrastructure and communities to minimise future disaster risks.

The chapter argued that such disaster risk mitigation requires socio-ecological and transformative approaches to resilience over engineering ones as systemic risks require systemic solutions. Thus, there has been much theorising about the best ways of changing the systemic social, economic and political barriers that weaken community capacities for mitigating hazard risks and their environmental drivers. The 2024 UNEP–ISC *Navigating New Horizons* report identified three categories of approaches to achieving this: collaborative governance, evidence-based policy making and a new social contract based upon shared values and improved social capital.

These approaches represent a significant meta-strategy. However, they also represent 'big picture' principles that tend to be ambiguous when it comes to actually implementing them. Thus, Vandenbroeck (2012) reminds us that achieving 'change in complex and contested settings remains a slippery mandate, beyond the reach of cookbook (or engineering) approaches' (p. 47). Thus, the final section of the chapter recommended breaking down the immensely entangled complexity of complex adaptive systems into a sequence of smaller, strategic 'tame problems' and applying some of operational strategies for resolving them – all the while treating all planned interventions in the system as experiments to be monitored and revised as the system's components and/or their interlinkages change as a result. These included: participatory learning and action, appreciative inquiry, community-based adaptation planning, strategic foresight, design thinking and scenario planning. All these strategies were identified as having the characteristics of systemic design. Integrating systems analysis with design thinking, systemic design represents a suite of principles for achieving action on and successful outcomes from the overriding approaches of collaborative governance, evidence-based policy making and a sharing of values and improved social capital. The chapter concluded with a list of seven iterative steps in implementing any systemic design strategy.

## Acknowledgements

The research for this paper was supported by Natural Hazards Research Australia.

## References

AIDR (Australian Institute for Disaster Resilience) (2020). AIDR, East Melbourne, Victoria, Australia. Available at: https://knowledge.aidr.org.au/media/7989/aidr_handbookcollection_communityengagementfordisasterresilience_2020.pdf (accessed 11 July 2025).

Alaimo, L. (2021) Complex systems and complex adaptive systems. In: Maggino, F. (ed.) *Encyclopedia of Quality of Life and Well-Being Research*. Springer Nature, Cham, Switzerland. DOI: 10.1007/978-3-319-69909-7_104659-1.

Brink, E. and Wamsler, C. (2018) Collaborative governance for climate change adaptation: Mapping citizen–municipality interactions. *Environmental Policy and Governance* 28, 82–97. DOI: 10.1002/eet.1795.

Bull-Kamanga, L., Diagne, K., Lavell, A., Leon, E., Lerise, F. *et al*. (2003) From everyday hazards to disasters: The accumulation of risk in urban areas. *Environment and Urbanisation* 15(1), 193–203.

Burger, A., Kennedy, W. and Crooks, A. (2021) Organizing theories for disasters into a complex adaptive system framework. *Urban Science* 5(3), 61. DOI: 10.3390/urbansci5030061.

Cinner, J. and Barnes, M. (2019) Social dimensions of resilience in social-ecological systems. *One Earth* 1(1), 51–56. DOI: 10.1016/j.oneear.2019.08.003.

Coetzee, C., Van Niekerk, D. and Raju, E. (2016) Disaster resilience and complex adaptive systems theory: Finding common grounds for risk reduction. *Disaster Prevention and Management* 25(2), 196–211. DOI: 10.1108/DPM-07-2015-0153.

CRED (Centre for Research on the Epidemiology of Disasters) (2025) *2024 Disasters in Numbers*. Available at: https://files.emdat.be/reports/2024_EMDAT_report.pdf (accessed 18 July 2025).

Cretney, R. (2014) Resilience for whom? Emerging critical geographies of socio-ecological resilience. *Geography Compass* 8/9, 627–640. DOI: 10.1111/gec3.1215.

Cutter, S. (2021) The changing nature of hazards and disaster risk in the Anthropocene. *Annals of the American Association of Geographers* 111(39), 819–827. DOI: 10.1080/24694452.2020.1744423.

De Roo, G., Yamu, C. and Zuidema, C. (eds) (2020) *Handbook on Planning and Complexity*. Edward Elgar Publishing, Cheltenham, UK.

Drezner, D. (2023) Are We Headed Toward a 'Polycrisis'? The Buzzword of the Moment, Explained. Vox. Com. Available at: https://www.vox.com/23572710/polycrisis-davos-history-climate-russia-ukraine-inflation (accessed 11 July 2025).

Edwards, F., Popartan, L. and Pettersen, I. (eds) (2023) *Urban Natures: Living the More-than-Human City*. Berghahn Books, New York.

Elgizawy, E. and Elewa, A. (2022) The impact of public participation on implementing landscape urbanism projects. In: Alberti, F., Amer, M., Mahgoub, Y., Gallo, P., Galderisi, A. *et al.* (eds) *Urban and Transit Planning*. Springer, Cham, Switzerland, pp. 139–148.

Eveleigh, N. (2023) What Is a Polycrisis, Why Is Everyone Talking About It and How Could It Affect Your Business? McGregor Boyall. Available at: https://www.mcgregor-boyall.com/resources/blog/what-is-a-polycrisis-how-could-it-affect-your-business/ (accessed 11 July 2025).

Fischer, J. and Riechers, M. (2019) A leverage points perspective on sustainability. *People and Nature* 1(1), 115–120. DOI: 10.1002/pan3.13.

Folke, C. (2016) Resilience (republished). *Ecology and Society* 21(4), 44. DOI: 10.5751/ES-09088-210444.

Gissing, A., Timms, M., Browninga, S., Crompton, R. and McAneney, J. (2022) Compound natural disasters in Australia: A historical analysis. *Environmental Hazards* 21(2), 159–173. DOI: 10.1080/17477891.2021.1932405.

Glaeser, E.L. (2022) Urban resilience. *Urban Studies* 59(1), 3–35. DOI: 10.1177/00420980211052230.

Goh, K. (2021) *Form and Flow: The Spatial Politics of Urban Resilience and Climate*. MIT Press, Cambridge, Massachusetts.

Hamdi, N. and Goethert, R. (1997) *Action Planning for Cities: A Guide to Community Practice*. Wiley, Oxford, UK.

Hassink, R. and Chu, H. (2025) Resilience. In: Warf, B. (ed.) *The Encyclopedia of Human Geography*. Springer, Cham, Switzerland. DOI: 10.1007/978-3-031-25900-5_253-1.

Head, B. (2022) *Wicked Problems in Public Policy*. Palgrave Pivot, Basingstoke, UK.

Hou, X., Li, R. and Song, Z. (2022) A bibliometric analysis of wicked problems: From single discipline to transdisciplinarity. *Fudan Journal of Humanities and Social* 299–329. DOI: 10.1007/s40647-022-00346-w.

Hurtado, P. and DeAngelis, J. (2022) The use of foresight and scenario planning in hazard mitigation and climate adaptation planning. *American Planning Association, PAS Memo* 113. Available at: https://www.planning.org/publications/document/9257226/ (accessed 11 July 2025).

Jones, H. (2014) Systemic design principles for complex social systems. In: Metcalf, G. (ed.) *Social Systems and Design*. Springer, Tokyo, pp. 91–128.

Kindon, S., Pain, R. and Kesby, M. (2007) *Participatory Action Research Approaches and Methods: Connecting People, Participation and Place*. Routledge, Abingdon, UK.

Lawrence, M. (2024) Polycrisis in the Anthropocene: An invitation to contributions and debates. *Global Sustainability* 7, e5. DOI: 10.1017/sus.2024.2.

Lawrence, M., Janzwood, S. and Homer-Dixon, T. (2022) What is a global polycrisis? And how is it different from a systemic risk? Technical Paper #2022-4. The Cascade Institute. Available at: https://cascadeinstitute.org/technical-paper/what-is-a-global-polycrisis/ (accessed 11 July 2025).

Lee, A. (2020) *Design by Resilience*. Springer, Cham, Switzerland.

Lee, K. (1993) *The Compass and the Gyroscope: Integrating Science and Politics for the Environment*. Island Press, Washington, DC.

Lennox, J. (2024) Simple systems and complex systems. In: *Robert Rosen and Relational System Theory: An Overview*. Springer, Cham, Switzerland, pp. 51–77.

Lenton, T., Rockström, J., Gaffney, O., Rahmstorf, S., Richardson, K. *et al.* (2019) Climate tipping points – too risky to bet against. *Nature* 575, 592–595. DOI: 10.1038/d41586-019-03595-0.

Levin, K., Cashore, B., Bernstein, S. and Auld, G. (2012) Overcoming the tragedy of super wicked problems: Constraining our future selves to ameliorate global climate change. *Policy Science* 45, 123–152. DOI: 10.1007/s11077-012-9151-0.

Liu, J., Dietz, T., Carpenter, S., Folke, C. and Trade, S. (2021) Coupled human and natural systems: The evolution and applications of an integrated framework. *Ambio* 50, 1778–1783. DOI: 10.1007/s13280-020-01488-5.

Matyas, D. and Pelling, M. (2014) Positioning resilience for 2015: The role of resistance, incremental adjustment and transformation in disaster risk management policy. *Disasters* 39(S1), s1–s18. DOI: 10.1111/disa.12107.

McNaught, M., Nalau, J., Hales, R., Pittaway, E., Handmer, J. *et al.* (2024) Innovation and deadlock in governing disasters and climate change collaboratively – Lessons from the Northern Rivers region of new South Wales, Australia. *International Journal of Disaster Risk Reduction* 105, 104366. DOI: 10.1016/j.ijdrr.2024.104366.

MCR2030 (n.d.) My City Is Getting Ready. Is Yours? UNDRR. Available at: https://mcr2030.undrr.org/ (accessed 11 July 2025).

Meyer, K. and Newman, P. (2020) *Planetary Accounting: Quantifying How to Live Within Planetary Limits at Different Scales of Human Activity*. Springer, Cham, Switzerland.

Miyagawa, S. and Ohta, S. (2022) Risk-based approach for VUCA world. In: Kanbara, S., Miyagawa, S. and Miyazaki, H. (eds) *Disaster Nursing, Primary Health Care and Communication in Uncertainty*. Springer, Cham, Switzerland, pp. 263–272.

Morin, E. and Kern, A. (1999) *Homeland Earth: A Manifesto for the New Millennium – Advances in Systems Theory, Complexity, and the Human Sciences*. Hampton Press, New York.

Napier, A. and Simister, N. (2017) *Participatory Learning and Action (PLA)*. INTRAC, Oxford, UK. Available at: https://www.intrac.org/wpcms/wp-content/uploads/2017/01/Participatory-learning-and-action.pdf (accessed 11 July 2025).

Nath, S. (2024) Mobilising transformative community-based climate change adaptation. *Urban Transformation* 6(1). DOI: 10.1186/s42854-023-00059-7.

Nelson, H. (2022) Design, Wicked Problems & Throwness. Accidental Vagrant. Available at: https://accidentalvagrant.blogspot.com/2012/10/design-wicked-problems-throwness.html (accessed 11 July 2025).

Nohra, C., Pereno, A. and Barbero, S. (2020) Systemic design for policy-making: Towards the next circular regions. *Sustainability12* 4494. DOI: 10.3390/su12114494.

Nunn, P., Klöck, C. and Duvat, V. (2021) Seawalls as maladaptations along island coasts. *Ocean & Coastal Management* 205, 105554. DOI: 10.1016/j.ocecoaman.2021.105554.

OCHA (UN Office for the Coordination of Humanitarian Affairs) (2024) Philippines: Six Tropical Cyclones Humanitarian Snapshot. Reliefweb. Available at: https://reliefweb.int/report/philippines/philippines-six-tropical-cyclones-humanitarian-snapshot-20-october-present-13-november-2024 (accessed 11 July 2025).

OECD (2025) *Strategic Foresight Toolkit for Resilient Public Policy: A Comprehensive Foresight Methodology to Support Sustainable and Future-Ready Public Policy*. OECD Publishing, Paris. Available at: https://www.oecd.org/content/dam/oecd/en/publications/reports/2025/01/foresight-toolkit-for-resilient-public-policy_9ad1cd60/bcdd9304-en.pdf (accessed 11 July 2025).

Pedgley, O. and Şener, B. (2024) Natural disaster readiness and response: Bringing designers, design thinking, and design innovation into the agenda. *She Ji: The Journal of Design, Economics, and Innovation* 10(1), 120–138. DOI: 10.1016/j.sheji.2024.04.001

Pequeño, A. and Ray, S. (2025) California Wildfire Updates: Palisades, Eaton Fires Near Full Containment. Forbes. Available at: https://www.forbes.com/sites/antoniopequenoiv/2025/01/31/california-wildfire-updates-hughes-fire-100-contained-as-palisades-eaton-fires-near-full-containment/ (accessed 11 July 2025).

Peters, B. (2017) What is so wicked about wicked problems? A conceptual analysis and a research program. *Policy and Society* 36(3), 385–396. DOI: 10.1080/14494035.2017.1361633.

Rayner, S. (2006) Wicked problems: Clumsy solutions – diagnoses and prescriptions for environmental ills. Jack Beale Memorial Lecture on Global Environment. ANSW Sydney. Available at: https://core.ac.uk/download/pdf/288283455.pdf (accessed 18 July 2025).

Riddell, G., Delden, H., Maier, H. and Zecchin, A. (2019) Exploratory scenario analysis for disaster risk reduction: Considering alternative pathways in disaster risk assessment. *International Journal of Disaster Risk Reduction* 39, 101230. DOI: 10.1016/j.ijdrr.2019.101230.

Riley, B., Willis, C., Holmes, B., Finegood, D., Best, A. *et al.* (2017) Systems thinking and dissemination and implementation research. In: Brownson, R., Colditz, G. and Proctor, E. (eds) *Dissemination and Implementation Research in Health: Translating Science to Practice*, 2nd ed. Oxford Academic Press, Oxford, UK.

Ripple, W., Wolf, C., Lenton, T., Gregg, J., Natali, S. *et al.* (2023) Many risky feedback loops amplify the need for climate action. *One Earth* 6(2), 86–89. DOI: 10.1016/j.oneear.2023.01.004.

Rittel, H. and Webber, M. (1973) Dilemmas in a general theory of planning. *Policy Sciences* 4, 155–169. DOI: 10.1007/BF01405730.

Roberts, N. (2000) Wicked problems and network approaches to resolution. *International Public Management Review* 1, 1–19. Available at: https://hdl.handle.net/10945/55330

Rockefeller Foundation (2024) 100 Resilient Cities. Available at: https://www.rockefellerfoundation.org/report/100-resilient-cities/ (accessed 18 July 2025).

Ryle, G. (2009/1949) *The Concept of Mind*. University of Chicago Press, Chicago, Illinois and Routledge, Abingdon, UK.

Sarkar-Swaisgood, M. and Dewi, M. (2020) *Developing cascading risk scenarios for complex disasters: pathways to manage risk and protect people*. Asia Pacific Disaster Resilience Network, Policy Study (4/2020). Available at: https://eciph.in/publication/developing-cascading-risk-scenarios-for-complex-disasters-pathways-to-manage-risk-and-protect-people/ (accessed 11 July 2025).

Scheffran, J. (2016) From a climate of complexity to sustainable peace: viability transformations in the anthropocene. In: Brauch, H., Oswald Spring, Ú., Grin, J. and Scheffran, J. (eds) *Handbook on Sustainability Transition and Sustainable Peace*. Springer, Cham, Switzerland, pp. 305–346.

UNEP (United Nations Environment Programme) (2024) *Navigating New Horizons: A Global Foresight Report on Planetary Health and Human Wellbeing*. UNEP, Nairobi. Available at: https://wedocs.unep.org/20.500.11822/45890 (accessed 11 July 2025).

Van Ael, K., Vandenbroeck, P., Jones, P. and Ryan, R. (2021) Systemic Design Toolkit. Available at: https://www.systemicdesigntoolkit.org/ (accessed 11 July 2025).

Vandenbroeck, P. (2012) *Working with Wicked Problems*. King Baudouin Foundation, Brussels. Available at: https://shiftn.com/_uploads_pdf/Wicked-Problems_KBF_2012-kopie.pdf (accessed 11 July 2025).

WEF (World Economic Forum) (2023a) World Economic Forum Annual Meeting. Available at: https://www.weforum.org/meetings/world-economic-forum-annual-meeting-2023 (accessed 11 July 2025).

WEF (World Economic Forum) (2023b) Meeting Overview: Cooperation in a Fragmented World. Available at: https://www.weforum.org/meetings/world-economic-forum-annual-meeting-2023/about/meeting-overview/ (accessed 11 July 2025).

WEF (World Economic Forum) (2023c) *Global Risks Report 2023*. Available at: https://www.weforum.org/publications/global-risks-report-2023/digest/ (accessed 11 July 2025).

Whitney, D. and Trosten-Bloom, A. (2010) *The Power of Appreciative Inquiry: A Practical Guide to Positive Change*, 2nd edn. Berrett-Koehler Publishers, Oakland, California.

Yang, N., Bose, M. and Horrigan, P. (2024) Landscape architecture as democratic practice: Learning from participatory methods and motivations in community-engaged design. *Landscape Research* 50(1), 58–73. DOI: 10.1080/01426397.2024.2409745.

Yarina, L. (2018) Your Sea Wall Won't Save You. *Places Journal*. Available at: https://placesjournal.org/article/your-sea-wall-wont-save-you/ (accessed 11 July 2025).

Zalasiewicz, J., Thomas, J., Waters, C., Turner, S. and Head, M. (2024) The meaning of the Anthropocene: Why it matters even without a formal geological definition? *Nature* 632, 980–983. DOI: 10.1038/d41586-024-02712-y.

# 5 Is Better Policy Enough? Governance of Urban Energy Transitions in the Global South

**Saul Roux[1]* and Clifford Shearing[2]**
[1]*Head: Environmental Strategy Implementation, City of Cape Town, South Africa;*
[2]*Professor Emeritus at Public Law, Faculty of Law University of Cape Town, South Africa and at the Centre for Criminology and Sociolegal Studies, University of Toronto, Canada; Honorary Professor, School of Regulation and Global Governance, The Australian National University, Canberra, Australia*

**Abstract**

This chapter analyses the conditions under which socio-technical systems, in this case electricity systems, transition to more sustainable configurations. It uses a study of the City of Cape Town (henceforth the City) to explore energy reconfigurations with a focus on the conditions under which systems change; potential pathways and typologies of change; and the governance of change. This investigation is situated within debates on socio-technical transitions in general and the multi-level perspective (MLP) in particular. The City's southern geography offered a lens though which to explore the ways in which technology interacts with a social context characterised by poverty and inequality. It explores themes related to contestation within socio-technical regimes; social contracts framing socio-technical regimes; nested multi-level governance; the politics of transitions; and the role of collective actors or networks and state actors in transitions. The research found that systemic tensions and contestation were prominent in relation to the incumbent electricity system. It demonstrated how regime reconfiguration based on environmental values competed with developmental values embedded within incumbent regime structures. This provides the basis for conceptualising socio-technical transitions as conflicts related to contested values. Overall, the chapter argues that competing value systems are integral in the co-evolutionary process of regime configuration and reconfigurations.

## Introduction

It is widely held that infrastructure transitions are urgently required to address intersecting challenges related to climate change, poverty and urbanisation – and the multiple crises of the Anthropocene. This chapter uses a study of the City of Cape Town (henceforth the City) to explore energy reconfigurations with a focus on conditions, pathways and governance. We situate our analysis within the context of debates related to the conditions that make reconfigurations possible; contestation within socio-technical regimes; nested multi-level governance; the politics of transitions; and the role of collective actors or networks and state actors in transitions (Berkhout *et al.*, 2004; Geels, 2004, 2010; Smith and Stirling, 2008).

*Corresponding author: saulroux@gmail.com

 *Climate Change and Risk Mitigation: Reducing Vulnerabilities and Enhancing Resilience* (ed. P. Barnes)
DOI: 10.1079/9781800624153.0005

The chapter draws on the findings of a study of the City's electricity distribution arrangements, conducted between 2012 and 2016. The City's southern geography offered a lens though which to explore the ways in which technology interacts with a social context characterised by poverty and inequality. Fieldwork was undertaken using a participatory, engaged and grounded theory approach (Layder, 1993). The research was conducted within a knowledge co-production (see, for example, Gibbons, 2000; May and Perry, 2006; Patel *et al.*, 2015; Greyling *et al.*, 2016; Djenontin and Meadow, 2018) context[1] that involved Roux, as a PhD student, spending 3 years in the City, embedded in the City's Energy and Climate Change Unit.

The chapter explores debates related to the conditions that make reconfigurations possible (Geels, 2002; Smith, 2007; Genus and Coles, 2008); contestation within socio-technical regimes (Genus and Coles, 2008; Lawhon and Murphy, 2011); social contracts framing socio-technical regimes; nested multi-level governance (Ostrom, 2010; Samuels and Samuels, 2018); the politics of transitions (Meadowcroft, 2009; Monstadt, 2009; Murphy, 2015); and the role of collective actors or networks (Monstadt, 2009; Castán Broto, 2015; Murphy, 2015) and state actors (Rotmans *et al.*, 2001) in transitions.

## Understanding Transitions

It is widely held that infrastructure transitions, such as energy infrastructures, are required to address intersecting challenges related to climate change, poverty and urbanisation. Further, such changes are required at a rate faster than typical long-term technological evolutions (Rip and Kemp, 1998; Geels and Schot, 2007). The immutable nature of large socio-technical systems becomes problematic when confronted with an urgent need for change (Kemp and Loorbach, 2006; Hodson, 2008). Over the past several decades, a field of research has arisen that aims to understand how more sustainable socio-technical configurations can be developed (Schot, 1992; Kemp *et al.*, 1998). This field of enquiry has become known as 'sustainable transition theory'. Various frameworks have gained prominence, including transition management (Kemp *et al.*, 1998; Rotmans *et al.*, 2001; Loorbach, 2010), strategic niche management (Kemp *et al.*, 1998; Smith, 2007) and technological innovation systems (Rip, 1995).

Recently, scholars have developed a 'multi-level perspective' (MLP) on socio-technical transitions (Geels, 2002; Geels and Schot, 2007; Smith and Grin, 2010) in order to analyse socio-technical systems and the conditions under which they change in response to wider social, political and economic dynamics (Rip, 1995; Rip and Kemp, 1998; Geels, 2004, 2010). According to the MLP, socio-technical transitions may occur in instances where established 'regimes' confront multiple systemic pressures that compel rapid and dramatic change (Rip and Kemp, 1998) when nascent technologies ('niche innovations') provide pathways for this (Geels, 2004). The MLP has proved useful for exploring the 'possibility of accelerating transitions' (Kemp, 1997; Kemp *et al.*, 1998).

This influential framing has been the subject of much debate. For example, scholars have criticised transition research in general, and the MLP in particular, for being overly functionalist (Shove and Walker, 2007; Genus and Coles, 2008) and for paying insufficient attention to power struggles (Smith and Stirling, 2008; Meadowcroft, 2009; Lawhon and Murphy, 2011) and for neglecting the role of collective actors, such as classes or social movements, in transitions (Monstadt, 2009; Castán Broto, 2015; Murphy, 2015). The MLP has also been criticised for failing to conceptualise institutional contradictions as drivers of change and a 'constitutive feature' of urban transitions (Castán Broto, 2015). For instance, Rip and Kemp (1998) have argued that technologies can have political effects built into them. Similarly, Lawhon and Murphy suggest that 'transition research must do more to analyse the ways in which artefacts are embedded into society and their consequent social impacts' (2011, p. 360). A cross-cutting theme in these criticisms is an argument for greater scrutiny of issues of power, resource control and social impacts of reconfiguration processes. In recognition of the value of conflict theories in responding to these critiques, transition studies have tended to integrate a range of concepts from these theories into their thinking (see, for example, Geels, 2002).

In this chapter we respond to these debates by exploring pathways of change in developmental context where issues of conflict and power relations are often paramount. To do so, we integrate ideas drawn from the sociology of technology, and in particular the work of Latour and collaborators. We deploy a constructivist lens to remain attuned to the 'specifics of development' and the relationship between 'technology and society' (Rice, 2013) that are crucial to understanding developments within Global South contexts.

In what follows, we pay particular attention to Geels and Schot's (2007) framing of transition while drawing inspiration from concepts such as splintered urbanism (Graham and Marvin, 2001; Jaglin, 2008; Jaglin, 2014) and differentiated infrastructure provision (Savage and Ward, 2003), and ideas developed within urban sociology and geography (Pieterse, 2011; Myers, 2014; Murphy, 2015). The overarching thread we use to integrate these themes is shifting political values.

## Historical Configuration and Reconfigurations in Cape Town

There have been four distinct periods in which widespread change in the City's electricity regime occurred. Each of these developments has been associated, and aligned, with shifts in value-sets specified in policy announcements.

### The initial configuration (1879–1912)

This initial period in the City's electricity history was characterised by experimentation and a fluid mix of municipal, private and hybrid public–private arrangements (Palser, 1995). For example, the first electric lights in Cape Town were installed through an agreement between the City, then under the government of the Cape Colony, and a private enterprise (Brush Electric Light and Power Company) to fit, maintain and guarantee electrical lighting. From the completion of the first private power station, in 1892, until the turn of the century, electricity was supplied and distributed by a mishmash of private and municipal entities and public–private partnerships.

During this period, the City played a central role in incubating and mainstreaming electricity 'niches' and innovations. For instance, the City led the conversion from direct current machines (dynamos) to alternating current machinery (alternators). This mainstreamed the use of alternators. Notably, the invention of the steam turbine in 1892, alongside the adoption of alternators, resulted in the use of turbo-alternator arrangements which revolutionised electricity generation. This 'revolution' was recognised by the City and it promptly adopted steam turbine technology in all new power stations and extensions. This 'early adoption' facilitated the future 'lock-in' of steam turbine technology. Thus, the initial configuration of the City's electricity regime emerged through collaborative processes between technology developers, the private sector and local government.

During this period, the electricity industry in Cape Town was regulated by a range of diverse and sometimes conflicting decrees, edicts and contracts. Despite regulatory uncertainty, these rules provided scope for innovation and experimentation. Although this configuration was shaped through relatively open values in terms of flexibility and collaboration, it laid the foundation for the emergence of segregated, exclusionary and discriminatory service delivery arrangements that lasted a century (Eberhard, 2010). Thus, the only interest groups that benefitted from this structure were corporates, the state and a minority of citizens, to the exclusion and detriment of all others.

### Reconfiguration for control and segregation (1912–1994)

The initial configuration was unravelled by the formation of the Union of South Africa (1910) and the rise of the apartheid government that reconfigured infrastructure systems from the top down. This new period was characterised by centralisation and monopolisation and a consolidation of power (Eberhard, 2010). In Cape Town, the early multiplicity of municipal, private and hybrid distributors was replaced by a centralised Electricity Supply Commission that consolidated the City-owned stations.

In 1913, the City's Council began acquiring generation and distribution assets of various private companies and successively dismantling private power stations. In parallel, between 1920 and 1922, the national government began to centralise the entire electricity industry. The Electricity Act of 1922 formally established the Electricity Supply Commission (ESCOM, today Eskom) that centralised control over electricity generation and distribution throughout South Africa, including control over all existing generation and distribution facilities.

By the mid-1960s, Eskom determined that all electricity supply to cities, including Cape Town, would be generated in the Limpopo and Mpumalanga power generation region – municipalities had to apply for licensing agreements to continue with existing generation and distribution facilities. During this period, several of the City's power stations were decommissioned, and future generation facilities in Cape Town were developed and financed solely by Eskom (Palser, 1995). This initiated the process of connecting Cape Town to an extensive South Africa-wide transmission system that was finalised by the end of 1970s. As a consequence, Cape Town became completely reliant on Eskom power and centralised coal-fired electricity generation.

The values that underpinned this regime configuration were aligned with the values of the apartheid regime. Energy policy was informed by two main threads – the provision of cheap energy for mining and primary industry and energy security for the apartheid state (Büscher, 2009; Eberhard, 2010).

## Reconfiguration for development and justice (1994 onwards)

The predominant dynamic that undermined and shifted the above configuration was the fall of apartheid that culminated in a democratically elected government being installed in 1994. The restructuring of the City's electricity regime during this period was brought about by a pincer movement comprised of bottom-up (political activism) and top-down (democratic change) pressures (Büscher, 2009; Krupa and Burch, 2011). In Cape Town, a network of activists, researchers and non-governmental organisations (NGOs), such as the Institute for Democracy in South Africa (IDASA) and the Energy and Development Research Centre (EDRC), were engaged in critiquing the incumbent electricity regime on the grounds of mass injustice. The process of reconfiguration of the City's electricity regime mirrored the social, political, legal, cultural and economic shifts that occurred during South Africa's democratic revolution.

With the collapse of apartheid, sweeping reform occurred in the energy sector driven by a variety of loosely aligned initiatives that sought to reconfigure access to electricity in ways that were underpinned by radically different values. By way of example, the White Paper on Energy Policy (DME, 1998) sought to redress historical inequalities by pursuing universal access to energy and ensuring accessible, affordable and reliable energy for poorer constituencies (Prasad, 2008). This new and now dominant value-set was embedded in a suite of new regulatory instruments, such as the Integrated National Electrification Programme, the Free Basic Electricity Policy (DME, 2003) and the Electricity Pricing Policy (DME, 2004).

Local governments were delegated widespread duties in the execution of these policies and programmes. This provided the political and regulatory framework for the City to restructure its electricity regime and to prioritise energy poverty alleviation, electrification and energy service provision. In 2000, a Cape Town metropolitan entity was established after almost a decade of local government restructuring. This resulted in the consolidation and rationalisation of five separate electricity departments (CCT, 2009a) within this new Unicity. A major value-based rationale for restructuring local governments was the extension of infrastructure services to the previously disadvantaged (CCT, 2009a; CCT, 2010b; CCT, 2011a).

Many City policy and planning documents highlight that the intended beneficiaries of this reconfiguration were citizens in general and the poor and marginalised in particular (CCT, 2022). However, what was retained in this new regime was centralisation, reliance on fossil fuels, spatial disconnects between generation and consumption and indirect support for South Africa's mineral-energy complex (Froestad *et al.*, 2018).

### Reconfiguration for environmental sustainability (1998 onwards)

The City is currently in the process of reconfiguring its electricity regime based on a further value shift, namely, a value-set that includes environmental values. A diverse set of stakeholders have facilitated this process – once again, this assemblage of supporting interests includes a loosely coupled network of NGOs, researchers and City-based innovators who have lobbied for over a decade for a lower-carbon, efficient and renewable urban electricity system and a more decentralised set of arrangements, motivated in large part by concerns about the ability of Eskom to deliver electricity to the City. Central to these delivery concerns has been escalating concern, since the mid-2000s, about 'state capture' (Chipkin *et al.*, 2018) and its consequences for Eskom's capacity to deliver electricity effectively and reliably to the City (CCT, 2023a, b). This assemblage of organisations was central in bringing energy security and climate change to the attention of City politicians and top management, with NGOs and researchers playing an important role as 'intermediaries' (Moss *et al.*, 2010) between separate City departments, officials and politicians and across spheres of municipal government.

Involvement in incubation programmes has also been central to the initiatives driven by this loose but influential alliance. A significant example was the City's participation in a Sustainable Energy for Environment and Development (SEED) programme, between 1998 and 2006, funded by the Danish International Development Agency (DANIDA) that played an influential role in kick-starting sustainable energy planning within the City. This DANIDA-supported processes culminated in a City Energy Strategies Conference in 2000, which eventually led to the adoption of the City's first Energy and Climate Strategy (CCT, 2006).

By 2008, the City's Integrated Development Plan (IDP), a primary municipal planning process required by national legislation, had formally recognised the importance of energy security and climate mitigation (CCT, 2009a). In the same year, Cape Town experienced major rolling blackouts and load-shedding as a result of a national supply crisis, associated with corruption, and its consequences for Eskom's ability to delivery electricity – a crisis that remains ongoing as we write. This crisis served to accelerate the acceptance of a value-set focused on energy security and energy justice, culminating in various iterations of climate mitigation policy, including the Energy and Climate Change Strategy (CCT, 2006), Integrated Metropolitan Environmental Policy (CCT, 2009b), Energy and Climate Action Plan (CCT, 2010a), Cape Town Energy 2040 (CCT, 2015), Climate Change Strategy (CCT, 2021) and most recently a 2050 Energy Strategy (CCT, 2023a). Alongside this, the City has reconfigured and refined institutional structures intended to implement these policies, including the establishment of the Sustainable Energy Markets Department, dedicated to coordinating a green and distributed energy transition (CCT, 2023a). Central to these developments has been an evolving shift in the value-set driving energy policy.

Once again, collaborations between the loosely coupled nodal assemblage (Holley and Shearing, 2017) of NGOs, the research community and City officials both raised awareness and prompted the development of policies which set out pathways intended to reconfigure the City's electricity regime (CCT, 2010a; Honig *et al.*, 2015).

Since 2012, Eskom has confronted a major financial and supply crisis that has accentuated the frequency of periods of rolling blackouts and load-shedding (Burton and Winkler, 2014). This continuing development provided the grounds for a renegotiation of energy configurations between national government and the City, motivated by a desire on the part of the City to establish greater security of supply. As we write, radically different roles and models for municipal supply are being considered and debated (DOE, 2019). The risk of an energy regime collapse, what Geels and Schot (2007) term a significant landscape shift, has shifted value-sets, and this, in turn, has opened up space for alternative configurations to take root.

## The Incumbent Electricity Regime

Cape Town confronts significant equity challenges related to electricity infrastructure. The City provides a high standard of electricity reticulation and associated services to affluent residential, commercial and industrial sectors. At the same time, many poor households and communities lack access to even basic energy services (SEA, 2014a). This pervasive energy inequity has its roots in colonial and apartheid legacies of deliberate value-driven exclusion and marginalisation (Büscher, 2009) with respect to electricity supply and energy service provision. The particular configuration of electricity generation and distribution infrastructure that was developed within the context of the supremist value-set that characterised the pre-1994 (the year of South Africa's first democratic elections) regime consistently produced marginalising social impacts (Büscher, 2009; Eberhard, 2010; Krupa and Burch, 2011).

As a consequence of these impacts, low-income households, unlike more affluent households, use mixed fuel sources that include, and often depend primarily upon, paraffin, candles and biomass. The use of these fuels has contributed directly to such issues as respiratory diseases and shack fires (Ea Energy Analyses, 2007). Lack of, or very limited, access to electricity, moreover, drives poor households to utilise illegal electricity connections that can and do result in electrocutions and fires (SEA, 2014a). Furthermore, as access to electrical energy is a significant driver of socio-economic development (Prasad, 2008), limited access to electricity, within impoverished communities, exacerbates poverty.

In South Africa, electricity is transmitted across long distances from the source of generation[2] to end users in cities (SACN, 2006; SEA, 2013) via distribution networks connected to the national grid. As a consequence, Cape Town is reliant upon centralised coal-fired power as a source of electricity, generated more than 1600 kilometres away, accessed via a national grid. This set of infrastructural arrangements has locked the City into a carbon-intensive electricity regime.

Within this context, the City supplies electricity within its boundaries by buying it in bulk from Eskom (the national utility), and then resells it to customers, after adding a surcharge (CCT, 2012b). The monies earned through these sales generate revenue that is used for a variety of purposes. These purposes include financing cost associated with electricity distribution within the City, including electrification, grid maintenance, metering installation and infrastructure roll-out (CCT, 2011a; CCT, 2012a). Notably, the City uses 'surpluses', after the above expenses have been paid, to significantly 'cross-subsidise' other municipal services.

The City has a historically acquired vested interest in both maintaining and indeed increasing sales of electricity acquired from a national electricity supply system dependent on fossil fuel-enabled generation. Furthermore, apart from sales, the City is committed, for reasons of equity, redistribution and energy security, to significantly increase the supply of electricity to its poor and marginalised communities. In response to historical and enduring energy poverty and marginalisation, the City's electricity regime has been configured to facilitate a bundle of 'values-driven' development objectives intended to redress apartheid infrastructure legacies (Eberhard, 2010). These equity- and justice-focused values include socio-economic development; equitable access to electricity and energy services; redistribution; affordable electricity provision; and social security for the urban poor. Given the legacy of an inherited electrical generation infrastructure, realising these values has come at a price – namely, support for a fossil fuel-based energy regime that is at odds with the City's explicit commitment to 'greening' their service delivery.

### Environmental values

A developmental value-set has, since 1994, been embedded into the fabric of the City's electricity regime. The newcomer in this assemblage of values has been the emergence of environmental values and the policy objectives associated with them. As we have indicated, there has been a loose advocacy network of nodes – drawn from across the public, private and NGO sectors – that have been active in seeking to get these values, and associated objectives, embedded in the City's policy and practices. For over two decades, a range of stakeholders have been actively engaged, both singularly and collectively, in initiatives intended to persuade both politicians and City officials to integrate environmental values into the City's developmental value-set. A principal terrain within which these initiatives have been pursued has been the decentralisation of energy generation via small-scale distributed generation. This has included options for municipalities to enter into power purchase agreements with renewable energy Independent Power Producers (IPPs), municipal-led energy generation, widespread adoption of energy efficiency objectives and the management of demand.

The emergence of an environmental value-set has led to significant internal fissures of contesting values. Given the significance of revenues from the sale of electricity and the impact on this revenue stream of decentralised generation, such as rooftop solar, as well as reductions in electrical consumption that environmental values encourage, it is no surprise that these value-driven fissures have played out in conflicts at political and administrative levels within the City as well as between competing networks (Janisch *et al.*, 2012; Trollip *et al.*, 2012; Laurent and Trollip, 2013).

Several City reports, including the City's Tariff Policy (CCT, 2023b), make it abundantly clear that electricity sales have formed the backbone of the City's revenue, and hence the capacity of the City to effectively deliver a range of municipal services, such as the provision of housing. These value conflicts have been exacerbated by a steady increase in the contribution of electricity sales to the City's total revenue.

The City's significant dependence on electricity revenue provides an institutional motivation to increase, rather than reduce, electricity consumption sourced from South Africa's coal-dependent national electricity grid, along with a disincentive to promote energy efficiency and consumers generating their own electricity from sustainable sources like rooftop solar (Laurent and Trollip, 2013; SEA, 2014b).

The ability of the City to provide affordable energy services to low-income households is dependent not only on its ability to maintain a continued revenue stream from the grid-based electricity consumption of mid- to high-income households and businesses – the entities most likely to engage in self-generation – but on its desire to increase the City's dependence on this high-carbon-related source of revenue, as the demand for services from low-income Capetonians increases (Trollip *et al.*, 2012; Laurent and Trollip, 2013; CCT, 2022). As a consequence, City officials, especially those committed to realising the dual objectives arising from an environmentally and equity-informed value-set, are caught between a proverbial 'rock and a hard place' of contesting values.

## Contesting values

The above discussion has outlined the systemic tensions between environmental and development values related to the City's electricity regime. This section further explores these tensions in the context of ongoing discussions to reconfigure the City's electricity regime. On the one hand, the City has committed to reconfiguring its electricity regime based on a range of development indicators. This constitutes an expansion and consolidation of the development values that currently underpin its electricity regime. This reconfiguration plan is entrenched in a range of City policies, including the IDP (CCT, 2022), Electricity Supply By-law (CCT, 2010b) and Tariff Policy (CCT, 2023b). The targets and indicators for this reconfiguration agenda include a commitment to increase access to grid electricity for residents living in informal settlements and 'backyard' dwellings on public land; the objectives of a backyard dwelling service project; and a commitment to ensuring access to energy services for poorer consumers through cross-subsidisation (CCT, 2023b) and the roll-out of an 'Energy access project' for low-income households (CCT, 2022). The indicators of this developmental agenda include, *inter alia*, energy poverty alleviation, cross-subsidisation, reducing inequality, universal access to electricity and energy services, affordability and provision of Free Basic Electricity (FBE).

At the same time, the City has committed to reconfiguring its electricity regime based on a range of environmental indicators. This plan is set out in a range of policies and plans, including the IDP (CCT, 2022), the Climate Change Strategy (CCT, 2021) and the 2050 Energy Strategy (CCT, 2023a). Commitments in relation to this agenda include diversifying supply away from coal-intensive Eskom towards IPPs utilising renewables, and enabling small-scale embedded generation; a City-initiated generation initiative; and an 'Energy demand response programme' (CCT, 2022).

Within this environmental and development policies context, a principal site of contestation has been the use of renewable energy, energy efficiency, demand-side management and behaviour change and the impact of these initiatives on the City's revenue stream and its capacity to cross-subsidise. There have also been contestations related to budget allocations; impacts of regulatory instruments that promote energy efficiency on development in general and on the poor in particular; and conversely, the increase in Cape Town's total electricity consumption and associated emissions that result from increased electrification and affordability for low-income households.

The development value-set and the agendas that express and support these values have driven sizable increases in access to affordable electricity and FBE, drawn from the national grid, by impoverished communities. These developments have directly sought to tackle the legacy of energy poverty for these communities that was a structural feature of the apartheid state. The continuation and expansion of this agenda is threatened by a low-carbon reconfiguration.

Within the City, the developmental value-set requires no lobbying to keep it on the City's agenda. Since 1994, this value-set has been taken for granted as a legitimate basis for driving policy and practice, as a value-set requiring action. This has not been true of the environmental value-set. While it is becoming increasingly recognised as a legitimate value-set that requires action, establishing its credentials and keeping it on the City's political agenda has been a significant and ongoing accomplishment and one that has required constant work. This work, and how it has been accomplished, is what we now consider.

## Governing Change within a Complex Socio-economic Landscape

In Cape Town there are well-established and active public–private urban policy networks comprised of state and non-state actors that have proved influential in shaping transitional processes by championing initiatives that support environmental values – initiatives that seek to challenge established patterns. Participation in these networks has resulted in a blurring of roles, attitudes and allegiances. Both state (e.g. the City line departments) and non-state actors (e.g. the NGO Sustainable Energy Africa) within these networks have spent over a decade establishing relationships across the City's administration that act as nodes of support.

These networks have employed a range of strategies and governance modalities to steer socio-technical change. Notably, they have actively engaged in 'niche incubation' (Kemp *et al.*, 1998; Smith, 2007) by supporting the nascent development and then mainstreaming 'green' innovations. This has been done in a number of ways, such as advocating for niche innovations in supply chain policies, and working to get these initiatives embedded in the City's Supply Chain Management Policy (CCT, 2017); supporting contracts with 'green'-focused energy service companies for municipal energy efficiency retrofits (CCT, 2011b); supporting niche innovations in behaviour change campaigns, such as the City's Electricity Savings Campaign; providing a supportive policy environment for green niche technologies through initiatives such as the Atlantis Greentech Special Economic Zone (CCT, 2021) and small-scale embedded generation (SSEG) feed-in tariffs (CCT, 2013). As highlighted by the report *Energy, Economics of Climate Change and the City of Cape Town* (CCT, 2012c), many of these nascent green technologies 'can only be viable with government and policy support', and these schemes thus seek to address the 'barriers to market penetration' (SALGA, 2013b) that niche technologies come up against (Smith, 2007).

A common strategy employed by networks supporting environmental value-sets has been to lobby national government for regulatory and policy reform in order to create top-down pressures and incentives for urban socio-technical change. City officials highlight that participation in these networks has enabled them to operate across traditional government scales.

These networks have established relationships with individuals across national government agencies, including the Department of Energy, the National Treasury, the National Energy Regulator and Eskom. Through these relationships, these networks have sought to influence national policy, using a range of mechanisms, including engagement in national Portfolio Committees; commenting on draft legislation, regulation and policy; and participation in national energy forums and meetings.

Related to the above, these networks have made use of umbrella organisations, including the South African City Network (SACN), the City Energy Support Unit, the South African Local Government Association (SALGA) and the Association of Municipal Electrical Utilities (AMEU), to consolidate and relay positions to national government, exchange knowledge, develop joint strategies and increase bargaining power against national government (SEA, 2009; SACN, 2013; SALGA, 2013a). These networks also engage directly with other South African cities in order to exchange knowledge and 'share lessons' (SALGA, 2013a). This has enabled the City to develop 'transition initiatives that have been taken up by the national government and then cascaded down to other cities' (Hodson and Marvin, 2010, p. 480).

These networks have also been successful in persuading City politicians to formally associate themselves with several transnational 'advocacy' networks (Lindseth, 2004). These include the C40 Cities Network, the Cities Climate Change Covenant, the carbonn Climate Cities Registry and the Clinton Climate Initiative (CCT, 2021). These international platforms have provided opportunities for resources, knowledge exchange and learning and have provided an avenue for pressuring colleagues to act in accordance with international commitments, targets and trends. A recent example is the City committing to the C40 Deadline 2020 programme.

## Re-framing Regimes as Value Systems and Social Contracts

The MLP, noted earlier, focuses on a range of elements that are solidified and entrenched within the structure of regimes including technology, regulation, markets and infrastructure (Geels, 2004). However, the MLP tends to overlook values as an element that both stabilises and is stabilised within socio-technical regimes. Further, the MLP conceptualises regime lock-in as driven primarily by macro processes that are ultimately value-free. Conversely, this chapter highlights that values both drive and are deeply entrenched in socio-technical regimes.

We have shown that the City's electricity regime is structured to facilitate a range of values, such as redistribution, social security, universal access to services in support of socio-economic development and affordable electricity for the poor. More than merely being embedded in the regime structure, it is evident that values are central to the way in which the City's incumbent electricity regime is configured. Further, these values both co-evolve alongside and exert a strong structuring force on other regime elements, including shaping the physical infrastructure of socio-technical regimes. This highlights that socio-technical regimes are both shaped by and underpinned by 'regime values'. Through this lens, socio-technical systems are (re)conceptualised as systems that integrate and house different values (and belief systems).

Further, the MLP regards the stabilisation of socio-technical regimes as driven by quasi-evolutionary dynamics that are, by and large, apolitical and detached from consensus-building and cooperation amongst social groupings (Smith, 2005; Shove and Walker, 2007; Scrase and Smith, 2009; Lawhon and Murphy, 2011). Conversely, the City's case highlights the presence of fragile but effective, albeit largely implicit, social compacts between diverse social groupings in relation to the municipal electricity regime. These compacts provide coherence to networks and provide a basis for collective action. Through this lens, transitions are seen as shaped by processes of ongoing renegotiations between different social groupings, made possible by integrating and unifying value-based social compacts, in response to landscape pressures that are forcing change.

## Socio-technical Reconfigurations as Value Tensions

The City's electricity regime is currently structured around a set of values that may broadly be defined as developmental values. However, as highlighted, these values have been, and are being, disrupted by green innovations and regime reconfigurations based on environmental value-sets. Thus, in the City's case, environmental values compete with developmental values embedded within the energy regime.

The chapter has explored how opposing regime values create tensions during regime reconfiguration processes, as entrenched regime values contest with new values, introduced by new alliances that challenge established alliances and their value-sets. These can lead to hybrid alliances and value-sets that seek to align what were previously regarded as conflicting values. These tensions are both systemic and manifest in reconfiguration policy agendas. This provides a conceptualisation of regimes as a set of contested and contesting values that compete with each other during reconfiguration processes. This analysis integrates MLP analyses with a Hegelian-Gramschian (Gramsci, 1971) perspective on transitions by conceptualising socio-technical regimes as systems produced and reproduced through contesting ideologies. It thus expands on debates by Shove and Walker (2007), Meadowcroft (2009), Monstadt (2009), Lawhon and Murphy (2011) and Murphy (2015), which consider the desirability of framing transitions as processes of contestation or as embedded in politics.

## Conditions Underpinning Regime Configurations and Reconfigurations

The above analysis yields a range of conceptual insights into regime change and the role and workings of the loosely coupled alliances that advocate for the incorporation of specific value-sets within socio-technical regimes. The first of these is that specific modes of thought, grounded in competing value-sets, are integrated into, and play a significant role in maintaining, socio-technical regimes. Further, the Cape Town case provides support for claims (see Kaghan and Bowker, 2001; Genus and Coles, 2008) that value-sets, while stable, are subject to challenges, and these challenges, if successful, can and do allow competing modes of thought to shape regulatory frames and the actions that they encourage and enable. As this happened, destabilising developments that received support, particularly regulatory support that tilted playing fields, gathered momentum and began to cascade.

The evidence also demonstrates how networks can be instrumental in facilitating the emergence of influential socio-technical assemblages. However, during processes of change, residues of previous regime values-sets may be absorbed within a reconfigured regime. As a consequence, over time, socio-technical regimes can be constituted through the layering of elements of different value-sets that become absorbed and stabilised. As this happens, regime reconfigurations emerge as reconfigured 'archaeological deposits' of diverse and sometimes competing value-sets. These findings support and extend discussions by Castán Broto *et al.* (2014) on examining institutional contradictions as drivers of change and central to transitions, as well as findings by Rip and Kemp (1998) that identified the political effects built into technologies. Further, the findings presented here support Lawhon and Murphy's call for further study of the ways in which 'artefacts are embedded into society and their consequent social impacts' (2011, p. 360).

Second, the Cape Town case demonstrates how value contests can, and do, enable shifts of institutionally supported sensibilities and motivations that in turn enable regime transitions. It provides evidence that shifts in values can, and do, lead to regime shifts. This analysis aligns with the ideas of conflict theorists (for example, Genus and Coles, 2008; Lawhon and Murphy, 2011) who adopt a Marxist-inspired approach to understanding the role of contestation in stimulating regime changes. This offers a perspective that complements and extends an understanding of the co-evolutionary process of regime formation and stability (see Geels, 2004). This reconceives regimes reconfigurations as grounded in systems of values that are shaped and reshaped by ongoing value struggles.

Third, the evidence on shifting alliances that networks enable cautions against relying exclusively on a structural account of transitions (Geels, 2004; Shove and Walker, 2007; Genus and Coles, 2008; Smith and Grin, 2010). What the Cape Town case highlights is the central role that activists, NGOs, communities and academics can, and do, play in fracturing regimes and creating fertile ground for change. This finding supports a conceptualisation of transitions that pays attention to contesting mentalities (Smith and Stirling, 2008) and the role of networks in building alliances of interests that come together to challenge established interest-based regimes.

Fourth, our analysis identified how a coalescence of disparate factors can constitute conditions that enable transitions. The findings we reviewed demonstrated how alliances of strange bedfellows – top-down[3] state-driven reconfigurations can work together with bottom-up/grass roots initiatives –emerge, especially at times of crisis, to open up novel spaces where new ways of thinking and new practices can take root. The evidence we canvased confirms how unexpected windows of opportunity (Kemp and Loorbach, 2006) – contingencies – provide opportunities for niche innovations to take hold (see Geels, 2002).

As we have demonstrated, disruptions or shocks (see Castán Broto *et al.*, 2014) at the landscape level can create space for innovative policy or influential initiatives that under normal circumstances (or during periods of regime stability) would not be entertained. Awareness and motivations (Honig *et al.*, 2015) regarding the value of change are heightened during periods of crisis. During these periods, conceived by the multi-level perspective as landscape shocks, existing socio-technical regimes may look less natural and in doing so may enable the emergence of a 'consciousness of purpose' (Gramsci, 1971: 598; see also Schön, 1979).[4]

## The Role of States in Reconfigurations

The results presented provide insights into the regime–niche relationship and the role of the state in reconfigurations. Due to the quasi-evolutionary underpinning of the MLP (Geels, 2010), the role of market-driven innovation is often conceptualised as central to facilitating transitions. However, this pays little attention to the actions of movements, civil society and the state. From the City's historical configuration and reconfiguration processes and the current (environmental) reconfiguration, it is evident that the relationship between the state (the City), niche innovations (innovation firms) and other actors (civil society) is more fluid than the 'rational choice' model offered by the multi-level perspective suggests (Geels, 2004; Meadowcroft, 2005, 2009; Murphy, 2015). The support the City provided to niche innovations goes beyond strategic niche management whereby nascent niche innovations are 'incubated' or protected from market competition through, *inter alia*, supportive regulation (Kemp *et al.*, 1998). In the City's case, in contrast, an alternative socio-technical regime, based on different values (and ideologies and belief systems) was envisioned through dialogue within a loose network. This network then nurtured a set of embryonic niche technologies in efforts to realise the envisioned regime. In this process, nodes in the City (state) played an important role in building a shared strategy for change across society; moulding an agenda for change; upscaling and mainstreaming niche innovations; creating learning opportunities; making landscape pressures visible; nurturing awareness; lobbying core regime members; and supporting transition trends already underway. These processes, rather than mere policy and regulatory support to incubate niche innovations, provided opportunities for firms to improve these innovations and hence compete with incumbent technologies. This underscores an iterative and mutually constituted process between the state and niche firms, which facilitated a 'breakthrough' of innovation.

This broadened understanding of the role of states in facilitating transitions is particularly relevant for exploring regimes in contexts where states play a role in the provision of network services. Our findings suggest that innovation studies (Schot and Geels, 2008) and transition management (Rotmans *et al.*, 2001; Loorbach, 2010) tend to over-emphasise the role of private niches in transitions. In the City administration, nodes and networks, with the specific function of facilitating regime reconfigurations, constituted what might be thought of as a 'public niche'. Our findings suggest that transitions can and do emerge from a combination of state intervention (or state innovation) and private sector innovation. In other words, collaboration between public niches and private niches can disrupt regimes. This confirms and expands on Rotmans *et al.* (2001), who highlight the role of state actors in steering transitions, initiating or directing transitions and as 'system builders' (Bijker, 1995) that support processes of aligning regulatory, technical, social and economic elements (Geels, 2004) of an envisioned socio-technical regime.

More broadly, sustainable transition research, including innovation and niche management studies, tend to regard niche (innovation) firms as the primary and indeed often as the sole generator of innovation. In contrast, our findings highlight the role of a broader set of actors and in particular the innovative impacts of not-for-profit organisations. This supports a broader conceptualisation of innovation that embraces a diversity of forms and builds on insights by Lawhon and Murphy (2011) and Genus and Coles (2008) that consider the contribution and interaction of diverse groups of actors in facilitating transitions.

## Conclusion

In this chapter we have explored the conditions under which systems change; potential pathways and typologies of change; and the governance of change. We have, throughout, highlighted that systemic tensions and contestation were prominent in relation to Cape Town's incumbent electricity system. We demonstrated how regime reconfiguration based on environmental values competed with developmental values embedded within incumbent regime structures. Further, we have considered historical reconfiguration processes in the City to explore the conditions that make change possible. We explored how crises open up space for regime reorganisation as competing value-sets are contested. This involved a discussion of the central role of activists, NGOs, communities and academics in fracturing regimes. We also identified the presence of public niches – or sections of the state with the specific function of facilitating change – and argued that collaboration between public niches and private niches disrupt regimes.

## Notes

1 Specifically the Mistra Urban Futures programme, funded by the Swedish International Development Cooperation Agency, located within the University of Cape Town's African Centre for Cities.

2 Mainly Mpumalanga.

3 Alternatively, command and control regulation.

4 Although it is recognized that crises as an opportunity for change is widely studied (Castán Broto *et al.*, 2014) in diverse fields, this study focused on Marxist ideas of crises due to the relationship between crises and contestation.

## References

Berkhout, F., Smith, A. and Stirling, A. (2004) Socio-technological regimes and transition contexts. In: Elzen, B., Geels, F. and Green, K. (eds) *System Innovation and the Transition to Sustainability: Theory, Evidence and Policy.* Edward Elgar Publishing, Cheltenham, UK, pp. 48–75.

Bijker, W. (1995) *Of Bicycles, Bakelites, and Bulbs: Toward a Theory of Socio-technical Change.* MIT Press, Cambridge.

Burton, J. and Winkler, H. (2014) *South Africa's Planned Coal Infrastructure Expansion: Drivers, Dynamics and Impacts on Greenhouse Gas Emissions.* Energy Research Centre, Research Report Series, University of Cape Town, Cape Town.

Büscher, B. (2009) Connecting political economies of energy in South Africa. *Energy Policy* 37(10), 3951–3958.

Castán Broto, V. (2015) Contradiction, intervention, and urban low carbon transitions. *Environment and Planning D: Society and Space* 33, 460–476.

Castán Broto, V., Glendinning, S., Dewberry, E., Walsh, C. and Powell, M. (2014) What can we learn about transitions for sustainability from infrastructure shocks? *Technological Forecasting & Social Change* 84, 186–196.

Chipkin, I., Swilling, M., Bhorat, H., Buthelezi, M., Duma, S. *et al.* (2018) *Shadow State: The Politics of State Capture*. Wits University Press, Johannesburg.

City of Cape Town (CCT) (2006) *Energy and Climate Change Strategy*. City of Cape Town, Cape Town.

City of Cape Town (CCT) (2009a) *5 Year Plan for Cape Town: Integrated Development Plan (IDP - 2007/8 – 2011/12) 2008/2009 Review*. City of Cape Town, Cape Town.

City of Cape Town (CCT) (2009b) *City of Cape Town Environmental Agenda 2009-2014: Integrated Metropolitan Environmental Policy for the City*. City of Cape Town, Cape Town.

City of Cape Town (CCT) (2010a) *Energy and Climate Change Action Plan for the City*. City of Cape Town, Cape Town.

City of Cape Town (CCT) (2010b) *Electricity Supply By-law*. City of Cape Town, Cape Town.

City of Cape Town (CCT) (2011a) *Electricity Department Business Plan 2012-2017*. City of Cape Town, Cape Town.

City of Cape Town (CCT) (2011b) *Report on Progress of Energy Efficiency Interventions in Council Buildings*. City of Cape Town, Cape Town.

City of Cape Town (CCT) (2012a) *Term of Office Five-Year Integrated Development Plan 2012-2017*. City of Cape Town, Cape Town.

City of Cape Town (CCT) (2012b) *City of Cape Town Budget for the Financial Period 2012/13 to 2014/15*. City of Cape Town, Cape Town.

City of Cape Town (CCT) (2012c) *Energy, Economics of Climate Change and the City of Cape Town: What Needs to Change*. Report to the Energy Committee, City of Cape Town, Cape Town.

City of Cape Town (CCT) (2013) *Report to the Energy Committee on Small-Scale Embedded Generation Electricity Tariff Structure*. City of Cape Town, Cape Town.

City of Cape Town (CCT) (2015) *Cape Town Energy 2040*. City of Cape Town, Cape Town.

City of Cape Town (CCT) (2017) *Supply Chain Management Policy (Incorporating Preferential Procurement)*. City of Cape Town, Cape Town.

City of Cape Town (CCT) (2021) *City of Cape Town Climate Change Strategy*. City of Cape Town, Cape Town.

City of Cape Town (CCT) (2022) *Five-Year Integrated Development Plan July 2022-June 2027*. City of Cape Town, Cape Town.

City of Cape Town (CCT) (2023a) *Our Shared Energy Future: The 2050 Energy Strategy for the City of Cape Town*. City of Cape Town, Cape Town.

City of Cape Town (CCT) (2023b) *Tariff Policy 2023/24*. City of Cape Town, Cape Town.

Department of Energy (DOE) (2019) *Integrated Resource Plan 2019 (IRP2019)*. Department of Energy, Pretoria.

Department of Minerals and Energy (DME) (1998) *White Paper on the Energy Policy of the Republic of South Africa*. Department of Minerals and Energy, Pretoria.

Department of Minerals and Energy (DME) (2003) *Electricity Basic Services Support Tariff (Free Basic Electricity) Policy*. Department of Minerals and Energy, Pretoria.

Department of Minerals and Energy (DME) (2004) *The Electricity Pricing Policy of the South African Electricity Industry*. Department of Minerals and Energy, Pretoria.

Djenontin, S. and Meadow, M. (2018) The art of co-production of knowledge in environmental sciences and management: Lessons from international practice. *Environmental Management* 61, 885–903.

Ea Energy Analyses. (2007) *Fighting Energy Poverty in Cape Town*. Ea Energy Analyses, Copenhagen.

Eberhard, A. (2010) Electricity supply. In: Jones, R. and Vivian, W. (eds) *South African Economy and Policy 1990-2000*. Manchester University Press, Manchester, UK.

Froestad, J., Nøkleberg, M., Shearing, C. and Trollip, H. (2018) South Africa's minerals-energy complex: Flows, regulation, governance and policing. In: Omorogbe, Y. and Order, A. (eds) *Ending Africa's Energy Deficit and the Law: Achieving Sustainable Energy for All in Africa*. Oxford University Press, Oxford, UK, pp. 308–343.

Geels, F. (2002) Technological transitions as evolutionary reconfiguration processes: A multi-level perspective and a case study. *Research Policy* 31(8–9), 1257–1274.

Geels, F. (2004) From sectoral systems of innovation to socio-technical systems: Insights about dynamics and change from sociology and institutional theory. *Research Policy* 33(6–7), 897–920.

Geels, F. (2010) Ontologies, socio-technical transitions (to sustainability), and the multi-level perspective. *Research Policy* 39(4), 495–510.

Geels, F. and Schot, J. (2007) Typology of sociotechnical transition pathways. *Research Policy* 36(3), 399–417.

Genus, A. and Coles, A. (2008) Rethinking the multi-level perspective of technological transitions. *Research Policy* 37(9), 1436–1445.

Gibbons, M. (2000) Mode 2 society and the emergence of context-sensitive science. *Science and Public Policy* 27(3), 159–163.

Graham, S. and Marvin, S. (2001) *Splintering Urbanism*. Routledge, Abingdon, UK.

Gramsci, A. (1971) *Selections from the Prison Notebooks of Antonio Gramsci*. In: Hoare, Q. and Smith, G. (eds). International Publishers, New York.

Greyling, S., Patel, Z. and Davison, A. (2016) Urban sustainability disjunctures in Cape Town: Learning the city from the inside and out. *Local Environment* 22(Suppl. 1), 52–65.

Hodson, M. (2008) Old industrial regions, technology and innovation: Tensions of obduracy and transformation. *Environment and Planning A* 40(5), 1057–1075.

Hodson, M. and Marvin, S. (2010) Can cities shape socio-technical transitions and how would we know if they were? *Research Policy* 39(4), 477–485.

Holley, C. and Shearing, C. (2017) Nodal Governance. In: Turner, B. (ed.) *The Wiley Blackwell Encyclopaedia of Social Theory*. Wiley, Oxford, UK, pp. 1–2.

Honig, M., Petersen, S., Shearing, C., Herbstein, T., Roux, S. *et al*. (2015) A conceptual framework to enable the changes required for a one-planet future. *Environmental Values* 24(5), 663–688.

Jaglin, S. (2008) Differentiating networked services in Cape Town: Echoes of splintering urbanism? *Geoforum* 39(6), 1897–1906.

Jaglin, S. (2014) Urban energy policies and the governance of multilevel issues in Cape Town. *Urban Studies* 51(7), 1394–1414.

Janisch, A., Euston-Brown, M. and Borchers, M. (2012) The potential impact of efficiency measures and distributed generation on municipal electricity revenue: Double whammies and death spirals. Technical paper. Sustainable Energy Africa, Cape Town. Available at: https://www.cityenergy.org.za/uploads/resource_23.pdf (accessed 4 August 2025).

Kaghan, W. and Bowker, G. (2001) Out of machine age? Complexity, socio-technical systems and actor network theory. *Journal of Engineering and Technology Management* 18(3–4), 253–269.

Kemp, R. (1997) *Environmental Policy and Technical Change*. Edward Elgar Publishing, Cheltenham, UK.

Kemp, R., Schot, J. and Hoogma, R. (1998) Regime shifts to sustainability through processes of niche formation: The approach of strategic niche management. *Technology Analysis and Strategic Management* 10(2), 175–195.

Kemp, R. and Loorbach, D. (2006) Transition management: A reflexive governance approach. In: Voss, J.P., Bauknecht, D. and Kemp, R. (eds) *Reflexive Governance and Sustainable Development*. Edward Elgar, Cheltenham, UK, pp. 103–130.

Krupa, J. and Burch, S. (2011) A new energy future for South Africa: The political ecology of South African renewable energy. *Energy Policy* 39(10), 6254–6261.

Laurent, L. and Trollip, M. (2013) *Residential, Photovoltaic Systems Roll-out in Cape Town - Scenario Modelling and Financial Details Assessment*. City of Cape Town, Cape Town.

Lawhon, M. and Murphy, J.T. (2011) Socio-technical regimes and sustainability transitions: Insights from political ecology. *Progress in Human Geography* 36(3), 354–378.

Layder, D. (1993) *New Strategies in Social Research: An Introduction and Guide*. Polity Press, Cambridge, UK.

Lindseth, G. (2004) The Cities for Climate Protection Campaign (CCPC) and the framing of local climate policy. *Local Environment* 9(4), 325–336.

Loorbach, D. (2010) Transition management for sustainable development: a prescriptive, complexity-based governance framework. *Governance: An International Journal of Policy, Administration and Institutions* 23(1), 161–183.

May, T. and Perry, B. (2006) Cities, knowledge and universities: Transformations in the image of the intangible. *Social Epistemology* 20(3–4), 259–282.

Meadowcroft, J. (2005) Environmental political economy, technological transitions and the state. *New Political Economy* 10(4), 479–498.

Meadowcroft, J. (2009) What about the politics? Sustainable development, transition management, and long term energy transitions. *Policy Science* 42(4), 323–340.

Monstadt, J. (2009) Conceptualizing the political ecology of urban infrastructures: Insights from technology and urban studies. *Environment and Planning A* 41(8), 1924–1942.

Moss, T., Guy, S., Marvin, S. and Medd, W. (2010) Intermediaries and the reconfiguration of urban infrastructures: An introduction. In: Guy, S., Marvin, S., Medd, W. and Moss, T. (eds) *Shaping Urban Infrastructures: Intermediaries and the Governance of Socio-Technical Networks*. Routledge, Abingdon, UK, pp. 1–14.

Murphy, J. (2015) Human geography and socio-technical transition studies: Promising intersections. *Environmental Innovations and Societal Transitions* 17, 71–89.

Myers, G. (2014) From expected to unexpected comparisons: Changing the flows of ideas about cities in a postcolonial urban world. *Singapore Journal of Tropical Geography* 35(1), 104–118.

Ostrom, E. (2010) Polycentric systems for coping with collective action and global environmental change. *Global Environmental Change* 20(4), 550–557.

Palser, D. (1995) *Lighting Up the Fairest Cape, 1895-1995: A Historical Record Commemorating the Centenary of the City of Cape Town Electricity Undertaking*. City of Cape Town, Cape Town.

Patel, Z., Greyling, S., Parnell, S. and Pirie, G. (2015) Co-producing urban knowledge: Experimenting with an alternative to 'best practice' for Cape Town, South Africa. *International Development Planning Review* 37(2), 187–203.

Pieterse, E. (2011) Recasting urban sustainability in the South. *Development* 54(3), 309–316.

Prasad, G. (2008) Energy sector reform, energy transitions and the poor in Africa. *Energy Policy* 36, 2806–2811.

Rice, J. (2013) Further beyond the Durkheimian problematic: Environmental sociology and the co-construction of the social and the natural. *Sociological Forum* 28(2), 236–260.

Rip, A. (1995) Introduction of new technology: Making use of recent insights from sociology and economics of technology. *Technology Analysis and Strategic Management* 7(4), 417–431.

Rip, A. and Kemp, R. (1998) Technological change. In: Rayner, S. and Malone, E.L. (eds) *Human Choice and Climate Change*. Battelle Press, Columbus, Ohio, pp. 327–399.

Rotmans, J., Kemp, R. and van Asselt, M. (2001) More evolution than revolution. *Foresight* 3(1), 1–17.

Samuels, A. and Samuels, A. (2018) Decarbonisation: The politics of transformation. In: Jordan, A., Huitema, D., van Asselt, H. and Forster, J. (eds) *Governing Climate Change*. Cambridge University Press, Cambridge, UK, pp. 15–34.

Savage, M. and Ward, K. (2003) *Urban Sociology, Capitalism and Modernity*. Palgrave Macmillan, London.

Schön, D. (1979) Generative metaphor: A perspective on problem-solving in social policy. In: Ortney, A. (ed.) *Metaphor and Thought*. Cambridge University Press, Cambridge, UK, pp. 254–283.

Schot, J. and Geels, F. (2008) Strategic niche management and sustainable innovation journeys: Theory, findings, research agenda, and policy. *Technology Analysis & Strategic Management* 20(5), 537–554.

Schot, J. (1992) The policy relevance of the quasi-evolutionary model: The case of stimulating clean technologies. In: Commbs, R., Saviotti, R. and Walsh, V. (eds) *Technological Change and Company Strategies: Economic and Sociological Perspectives*. Academic Press, London, pp. 185–200.

Scrase, I. and Smith, A. (2009) The (non)politics of managing low carbon sociotechnical transitions. *Environmental Politics* 18(5), 707–726.

Shove, E. and Walker, G. (2007) CAUTION! Transitions ahead: Politics, practice and sustainable transition management. *Environment and Planning A* 39(4), 763–770.

Smith, A. (2005) The alternative technology movement: An analysis of its framing and negotiations of technology development. *Human Ecology Review* 12(2), 106–119.

Smith, A. (2007) Translating sustainabilities between green niches and sociotechnical regimes. *Technology Analysis and Strategic Management* 19(4), 427–450.

Smith, A. and Stirling, A. (2008) Social-ecological resilience and sociotechnical transitions: critical issues for sustainability governance [*Working Paper 8*]. STEPS Centre, Brighton, UK.

Smith, A. and Grin, C. (2010) Innovation studies and sustainability transitions: The allure of the multi-level perspective and its challenges. *Research Policy* 39(4), 435–448.

South African Cities Network (SACN) (2006) *State of the Cities Report 2006*. South African Cities Network, Johannesburg.

South African Cities Network (SACN) (2013) *Consolidation of Lessons Learnt for EE and RE Initiatives within Cities, Development of a Roadmap for Future Uptake*. South African Cities Network, Johannesburg.

South African Local Government Association (SALGA) (2013a) *Local Government Energy Efficiency and Renewable Energy Strategy. Status Quo Report*. South African Local Government Association, Pretoria.

South African Local Government Association (SALGA) (2013b) *Regulatory and Policy Obstacles to the Adoption of Renewable Energy and Energy Efficiency Technologies by Municipalities*. South African Local Government Association, Pretoria.

Sustainable Energy Africa (SEA) (2009) *How to Implement Renewable Energy and Energy Efficiency Options: Support for South African Local Government*. Sustainable Energy Africa, Cape Town.

Sustainable Energy Africa (SEA) (2013) *Energy and Urbanisation in South Africa: Context Report and Literature Review*. Sustainable Energy Africa, Cape Town.

Sustainable Energy Africa (SEA) (2014a) *Tackling Urban Energy Poverty in South Africa*. Sustainable Energy Africa, Cape Town.

Sustainable Energy Africa (SEA) (2014b) *Case Study: Small-Scale Embedded Generation in South African Municipalities*. Sustainable Energy Africa, Cape Town.

Trollip, H., Walsh, V., Mahomed, S. and Jones, B. (2012) Potential impact on municipal revenue of small scale own generation and energy efficiency. In: *Conference paper presented at the South African Economic Regulators Conference, 21–22 August 2012*, National Energy Regulator of South Africa, Pretoria.

# 6 Hotter and Damper: Dealing with the Environmental Health Impacts of Changing Weather and Climate

**Enembe O. Okokon[1,2,3]*, Michael Tong[2] and Sotiris Vardoulakis[1]**
*[1]HEAL Global Research Centre, Health Research Institute, University of Canberra, Australia; [2]National Centre for Epidemiology and Population Health, Australian National University, Australia; [3]University of Calabar, Calabar, Nigeria.*

**Abstract**

Anthropogenic activities that emit greenhouse gases have intensified global warming. The daily mean and maximum temperatures and the frequency of extreme heat events have increased in some regions of the world. These changes are greater in tropical regions of the world, in summer, and in urban areas. Rising temperatures enhance the moisture-carrying capacity of air, thereby increasing humidity and heavy precipitation. As the intensity of precipitation events increase, storms and flood events become more frequent. Rising temperatures magnify inequities relative to energy consumption. Periods of high heat coincide with peak electricity consumption because of the compelling need for energy-driven indoor cooling. Affluent populations can afford the energy expenditure required to modulate the indoor microclimate, but the lean resources of the less-affluent cannot accommodate the additional cost of electricity during peak heat making them more vulnerable to the impacts of extreme heat. Lower-income countries are more prone to similar energy injustices. With suboptimal evaporative cooling in conditions of humid heat, populations become more susceptible to heat stress, poorer perceptions of wellbeing, and diseases affecting multiple organ systems like the heart, kidneys, lungs, and liver. Humid heat can also exacerbate pre-existing morbidities, while floods can damage property, destroy livelihoods, cause bodily harm, and interrupt municipal services. Adaptation to extremes of heat and precipitation is multifaceted and requires responsibility from the government, private citizens, and various stakeholders. Adaptive solutions must be well integrated embedded within the community, and centrally coordinated. These solutions include infrastructural planning, development, and regulation; targeted health and municipal services; responsive policy frameworks; and the provision of financial and relief support.

## Introduction

Global warming is a now conversant phenomenon that is affecting the atmosphere, land surfaces and ocean bodies around the world. According to the Intergovernmental Panel on Climate Change (IPCC) Sixth Assessment Report, the global average surface temperature has increased by 0.85°C between 1880 and 2012 (Whitmee *et al.*, 2015). In addition, the maximum global and regional temperatures have increased since the 1950s (Ebi *et al.*, 2021). A recognizable consequence of the warming trend is the increasing global and regional incidences of temperature extremes, such as heatwaves. The moisture-carrying capacity of the air

*Corresponding author: Enembe.Okokon@canberra.edu.au

 *Climate Change and Risk Mitigation: Reducing Vulnerabilities and Enhancing Resilience* (ed. P. Barnes)
DOI: 10.1079/9781800624153.0006

expands when air is heated (Trenberth, 1999). A warming atmosphere with more moisture generates more intense precipitation, causing floods (Ahern *et al.*, 2005). This concomitant increase in temperature and humidity is demonstrated by observed increases in dew-point temperature and specific humidity between 1951 and 1994 (Wang and Gaffen, 2001). Similar trends have also been recorded for the wet-bulb temperature (Raymond *et al.*, 2020), which has considerable implications for the human sensation of thermal comfort and the rising incidence of heat strain in different parts of the world (Raymond *et al.*, 2020). The combination of climate variability and long-term warming trends can generate extreme forest fires in certain regions during dry and hot conditions (Abram *et al.*, 2021).

With diminishing thermal comfort, human cooling needs have increased in many regions. Notably, a thermoneutral environment is necessary for crucial physiological functions such as sleep (Minor *et al.*, 2022). With rising night-time temperatures in many parts of the world, particularly in urban areas, indoor thermal comfort declines and the thermal burden on local populations becomes less sufferable. Accordingly, sleep is strained and inadequate for the physiologic requirement. Sleep insufficiency is a harbinger of disease and health conditions and a constraint on human wellbeing and productivity (Wang *et al.*, 2023).

Efforts to promote indoor comfort often culminate in substantial household expenditure on technologies which create desirable indoor environmental conditions microclimate. The frequent use of these technologies can only be supported with an energy budget which exerts further pressure on the household income (Walker and Day, 2012). Some of these technologies and the cost of operating them are beyond the reach of the economically underprivileged, for example those living in social housing. Many of these households occupy buildings which were constructed without deliberate designs for passive cooling (Londgen *et al.* 2022). It follows that an increased thermal burden not only hinders the health development of people, it also magnifies the energy security injustices that exist between people in different socio-economic strata. This is true for rural and urban dwellers (Oleson *et al.*, 2015).

Urban areas are generally warmer than rural areas because of architectural and planning features (for example, the lack of green space) that generate and retain heat in cities (Voogt, 2002). This urban heat island effect intensifies the impact of heatwaves on the population. The potential for heat-related illness is therefore higher in urban areas. These illnesses may be due to intense exposures which cause heat cramps, heat exhaustion, heat oedema, heat syncope and heatstroke (Borden and Cutter, 2008). These disorders are generally acute. Recurrent exposures can predispose to or exacerbate chronic diseases of the respiratory, cardiovascular, renal and nervous systems (Chambers *et al.*, 2023).

An adaptive response to increasing environmental temperature and humidity in conditions of climate change requires a multi-pronged approach. This approach should harness resources and expertise from different sectors. Stakeholder participation of government departments, health services, academia, the private sector, civil society and community groups should be coordinated to establish a synergy that is appropriate for approximating adaptation objectives at the tactical, strategic and operational levels. Adaptive solutions may include reimagining the morphology of cities, prioritising passive-cooling designs of buildings, providing health-heat warning systems and relevant health education before and during extreme heat events, and establishing cooling centres in cities. Data should be collected to evaluate the effectiveness of implemented adaptation strategies, so that the findings from previous efforts can be applied to improve future adaptation planning (Dwyer *et al.*, 2022).

## Shifting Earth Systems Drive Climate Change

Since the dawn of the industrial era, human dependence on fossil fuels to support industrial machines and transportation has steadily increased. The human population has increased by more than tenfold in the past three centuries. Consequently, there has been a substantial rise in the demand for food, goods and services. The production processes that are geared to match these demands inherently depend on activities that release greenhouse gases (GHGs) into the atmosphere. Vast slews of tropical forests and natural vegetation have been depleted because of human activities, population expansion and rapid urbanisation. These forests act as natural sinks to atmospheric carbon dioxide, which is necessary for photosynthesis. The concentration of carbon dioxide in the atmosphere has increased by more than 40% since the 18th century (IPCC, 2023). This era has been designated the Anthropocene, and it marks the period when human activity began to leave an indelible imprint on Earth's systems, causing substantial change to the planetary balance (Steffen *et al.*, 2011).

## The energy budget of the Earth: The failing balancing act

The energy budget of the Earth is maintained by an outward reflection of radiant energy, equivalent in magnitude to energy arriving from the sun. When long-wave solar radiation impacts the Earth from the sun, some of it gets immediately reflected off the clouds back into space; the rest is reflected off ocean bodies, deserts and aerosols in the atmosphere. This reflective property is known as the albedo. The Earth reflects incoming energy as infrared radiation. GHGs in the atmosphere absorb some of the infrared radiation and impede its outward trajectory to space. These gases emit some of the infrared energy towards the lower atmosphere and back towards the Earth, thereby increasing the surface temperature of the Earth. The abundance of GHGs in the atmosphere has intensified global warming in the Anthropocene. The global mean surface temperature has increased by 0.85°C between 1880 and 2012 (IPCC, 2023), and is projected to further increase by 2.6–4.8°C by the end of the 21st century (Whitmee *et al.*, 2015). Increases in the mean and extremes of global and regional temperatures have been observed since 1950, but the scale of the increase varies across different regions.

## Temperature extremes and heatwaves

Climatic trends have shown a statistically significant increase in the frequency of warm days, while cold nights are relatively fewer if compared with historical patterns (Seneviratne *et al.*, 2014). An analysis of 217 cities around the world has shown a significant increase in extreme hot days in about half of these cities, and hot nights have increased in 65% of cities (Horton *et al.*, 2016). Noticeably, there has been a more widely observed trend of greater night-time warming compared to daytime, but in regions where daytime warming exceeds night-time warming, the overall daytime warming is twice that experienced elsewhere (Cox *et al.*, 2020).

In addition to mean surface temperatures, maximum global and regional temperatures have also increased (Ebi *et al.*, 2021). As of 2024, the warmest year recorded using standardised measurements in 174 years was 2023. The global composite near-surface mean temperature for land and ocean was 1.45 ± 0.12°C higher than pre-industrial levels (WMO, 2024). However, the World Meteorological Organization declared 2024 as the warmest year on record with temperature exceeding 1.55°C above pre-industrial level (WMO, 2025), indicating a possible step-wise upward trajectory in surface temperature in recent years. The temperature near the land surface has increased at twice the rate for ocean surface temperatures (Solomon *et al.*, 2007). Daily maximum temperatures over land have been increasing in value and frequency of occurrence, coinciding with a gradual increase in the daily mean temperatures. It would seem that the increase in extreme temperature outpaces the rise in mean temperature (Horton *et al.*, 2016). Extreme and prolonged heatwaves have seen a tenfold increase in some regions of the world, notably parts of Europe, with a corresponding fivefold increase in global average (Coumou *et al.*, 2013). Extensive model simulations suggest that the return time for extreme heat in Europe has decreased from thousands of years to tens or hundreds of years (Christidis *et al.*, 2015). Warming rates and extremes will not be identical across the Earth but will differentiate regionally, depending on the local meteorological patterns (IPCC, 2023). A dreadful consequence of climate variability and extended warming trends is the increased risk of extreme forest fires during dry and hot conditions, like those experienced during the Australian Black Summer of 2019–2020 (Abram *et al.*, 2021).

## Heat tolerance and thermoregulation in extreme conditions

The human biology is unprepared for the heat-related problems that will emerge in a future marked by a historically unrivalled frequency of extreme heat events (Hanna and Tait, 2015). Frequent high temperatures will be an unrivalled health risk factor for exposed populations. It is logical to conceive that the ensuing thermal environment will test the biological limits of thermal tolerance – the range of body temperatures within which the physiological functions supporting life can exist. The optimal core body temperature conducive to appropriate physiological function and healthy life is 36.8 ± 0.5°C (Havenith, 2003; Hanna and Tait, 2015). The body homeostasis is distorted in situations where thermal discomfort persists. At very high temperatures, there will be a breakdown of physiological thermoregulation. When the core body temperature exceeds 38°C, the capacity of the body to adjust starts to fail. With a further increase in the core temperature, cellular death and multi-organ failure are likely to ensue (Raymond *et al.*, 2020). It has been observed that with every 1°C rise in ambient air temperature, all-cause morbidity and mortality increases by about 3% in regions of the world with annual mean temperature of 26°C (Guo *et al.*, 2017).

## The mechanism of thermoregulation

An understanding of how the body regulates heat should preface the discussion on the multiplier influence of co-exposure to heat and humidity on human health. There are two components of body temperature, namely, the core temperature – the temperature of structures and viscera that lie deep within the human body – and the peripheral or skin temperature. Human thermoregulation aims to maintain a reasonably constant deep body temperature that is appropriate for biological activity. The body modulates and sustains this internal environment by maintaining a state of balance, which safeguards the parity of heat gained and heat lost to the environment (Ebi *et al.*, 2021). The skin temperature is about 4°C lower than the core temperature; this gradient facilitates the outward transfer of heat from the body to the surrounding environment (Raymond *et al.*, 2020). Heat loss is affected by the dilatation of peripheral arteries to enable increased blood flow to the skin. Enhanced peripheral blood flow facilitates sweating. Skin moisture facilitates the loss of the latent heat of vaporisation needed to evaporate sweat from the surrounding skin, thereby cooling the body periphery in the process.

## Heat extremes and atmospheric moisture

At sustained wet-bulb temperatures exceeding 35°C, the body is in a state of hyperthermia (Sherwood and Huber, 2010). Hyperthermia is defined as an increase in core body temperature due to heat gain that is sufficient to exhaust the body's ability to lose heat (Hanna and Tait, 2015). Hyperthermia is more common when the wet-bulb temperature reaches and exceeds 35°C. Extreme wet-bulb temperatures that approach the 35°C survivability limit are occurring more frequently in the context of climate change (Raymond *et al.*, 2020).

Raised dry-bulb temperature causes the air to expand and enhances the moisture carrying capacity of the atmosphere. Rising surface (ambient) temperatures also potentiate evaporative moisture loss from the Earth's surface. The atmospheric moisture content expands as a corollary of these events (Trenberth, 1999). An augmented evaporative water loss depends on the local availability of sizable water bodies as well as land-based moisture sources, which let off higher volumes of water vapour when subjected to an intensifying radiative forcing (Trenberth, 1999). The warm air expands, and receives and retains the moisture until the point of saturation. When the carrying capacity at a given temperature and pressure is breached, precipitation invariably takes place. Data from 178 weather stations located in all parts of the USA (Robinson, 2000) and 196 stations located in four regions of China (Wang and Gaffen, 2001) indicate a rising trend of dew-point temperature and specific humidity, respectively, between 1951 and 1994.

Dry-bulb temperature, which measures sensible heat, is the temperature of ambient air measured by a thermometer unaffected by moisture or direct radiant heat from the sun. It represents the heat content of a parcel of air. In contrast, wet-bulb temperature is the temperature of the air that cools to saturation by water evaporating into it. This metric denotes a physical combination of heat and humidity. The wet-bulb temperature exceeds dew-point temperature, but saturation is possible at this temperature because of the moisture provided by evaporating air (Fischer and Knutti, 2015). The wet-bulb temperature of unsaturated air is always lower than the dry-bulb temperature, but both temperatures are equivalent at 100% relative humidity.

Wet-bulb temperature is of practical utility in evaluating heat stress relative to the air temperature. This is because when the wet-bulb temperature is high, evaporative cooling is less effective (Fischer and Knutti, 2015). Sweat evaporates if more moisture is present on the skin than in the air. If the moisture content of air surpasses skin moisture, sweat cannot be evaporated (Havenith, 2003). To this extent, humidity has been suggested to reinforce the mechanism by which atmospheric temperature changes affect human health (Barreca, 2012). However, the combined impact of heat and humidity on human health and function may be additive. Humidity magnifies heat stress by interfering with the body's ability to cool itself through sweating (Barreca, 2012). Therefore, it is expected that in hot and humid places, the incidences of cardiovascular and respiratory emergencies and mortality are likely to increase in the context of climate change (Ebi *et al.*, 2021).

As global warming intensifies, weather stations around the world have recorded wet-bulb temperatures exceeding 33°C more frequently in recent years. Raymond *et al.* (2020) reanalysed weather station data spanning the last four decades and reported a rapid increase in extreme wet-bulb temperature. Their findings highlight that on both land and sea, there is an increased incidence of wet-bulb temperatures that approach the survivability limit of 35°C in the subtropics. Humid heat will, therefore, be a manifest topical stressor requiring collaboration among public health specialists, social scientists, meteorologists and policy makers to address the health challenges set to emerge in a progressively warmer physical environment (Raymond *et al.*, 2020).

The recognition of similar atmospheric trends for heat and humidity, and the looming health consequences of the combined exposure, has prompted the development of heat-humidity indices, which cater for a more adequate characterisation of the atmospheric conditions affecting human thermoregulation and biological energy balance (Matthews, 2018). Environmental scientists have developed multiple indices that represent different aspects of human–environment heat exchange. These indices are relevant to different epidemiological research objectives (Davis *et al.*, 2016). Some of these indices can be instrumentally measured, while others are derived using mathematical equations and computational methods. Some of these heat stress indices are:

- The wet-bulb globe temperature (WBGT): This is a weighted average of the dry-bulb, wet-bulb and mean radiant (globe) temperatures. It has been widely used in situations pertinent to workplace safety, athletics and the military (Budd, 2008). It is derived as follows:

  $$WBGT = 0.7T_W + 0.2T_g + 0.1T_a$$

  where, $T_w$ is the wet-bulb temperature, $T_g$ is the globe temperature, and $T_a$ is the dry-bulb temperature.

- The temperature–humidity index (THI): This is also known as the discomfort index (Thom, 1959). It is based on the notation:

  $$THI = 0.4\left(T + T_W\right) + 15$$

  where $T$ is the ambient temperature measured in the shade (dry-bulb temperature), and $T_w$ is the wet-bulb temperature.
- The humidity index or humidex (*Hx*): This is used by Canadian meteorologist to represent the sensory experience of people due to the combined exposure to heat and humidity (Masterton and Richardson, 1979). It is derived as:

  $$Hx = T + 0.5555\left(e - 10\right)$$

  where, $T$ is the air temperature in degrees Celsius,

  $$e = \left\{6.11 \times exp\left[5417.7530\left(\frac{1}{273.15} - \frac{1}{273.15+T_{dew}}\right)\right]\right\}$$

- Apparent temperature, also known as 'felt temperature': This is the perception of temperature that combines the effect of dry-bulb temperature, humidity, wind speed and solar radiance. It can be derived in different ways, but the original version (Steadman, 1994) is as follows:

$$AT = T_a + 0.33vp - 0.70ws - 4.00$$

where, $T_a$ is the dry-bulb temperature, $vp$ is the vapour pressure

$vp = rh \times 6.105 \times e^{\left(\frac{17.27 \times T_a}{237.7+T_a}\right)}$, and $ws$ is the wind speed

Further indices have been developed, such as the discomfort index, the environmental stress index, corrected effective temperature, the heat index and others, which take into account human and physical environmental parameters such as age, human activity level, local environmental conditions, clothing, awareness and food consumption (Rachid and Qureshi, 2023).

The sensitivity of each index to changes in the underlying parameters varies distinctly from the others (Rachid and Qureshi, 2023). It can be daunting to make a selection of an appropriate indicator from the vast array of over 40 indices. Davis *et al.* (2016) therefore argued that a researcher's index of choice should be guided by the specific objective of the research and how humidity fundamentally relates to the process or condition of interest. For research or policy objectives that pertain to human comfort and evaporative cooling in heat stressful environments, the authors recommended apparent temperature, the temperature–humidity index, the humidex and wet-bulb temperature (Davis *et al.*, 2016). This view point aligns with the early position of Macpherson (1962), who maintained that the index of choice may be selected to assess the impact of a proposed remedial measure.

### Heat, precipitation and flood events

In climate change conditions, the increase in temperature expands the water-carrying capacity of the atmosphere and elevates surface-moisture loss through evapotranspiration. Increased water vapour enhances the potential for heavy precipitation (rainfall or snowfall) at saturation. Therefore, heavy downpour tends to occur with greater intensity and frequency in many regions of the world (Nikolowski *et al.*, 2013). Climate change also enables variability in the pattern of atmospheric circulation, which imposes changes in precipitation locally and regionally. These changes manifest in the location, type (rain or snow), amount, frequency, intensity and duration of precipitation (Trenberth, 2011). Changing precipitation patterns also imply an occurrence of drought in some places in contrast to heavy precipitation in others. However, a discussion of drought is outside the scope of this chapter (Pizzorni, *et al.*, 2024).

The changing precipitation does not translate to an increase in mean precipitation, but rather, torrential downpour occurs more frequently, even as the total volume of rain has declined over the years, thus the tendency to 'storm' rather than 'rain' (Trenberth, 2011). The altered patterns of precipitation tend to occur in association with sea-level rise (due to melting ice caps), cyclones and hurricanes as the sea surface temperature rises (Ahern *et al.*, 2005), compounding precursors for flood events. In low-lying places, these floods can occur very rapidly, in which instance they are known as 'flash floods'. Flash floods occur when runoff from torrential rainfall generates a sizeable body of water which inundates typically dry places or causes existing water bodies to overflow. Frequent flood events tend to occur in places where urban development and construction encroach on floodplains. Similarly, the reduction in green space degrades stormwater runoff mitigation (Vardoulakis *et al.*, 2015). Floods classification can be derived from the causative factors (heavy rainfall, rising tides, the breach of structures such as dams) or the nature of the flood (speed of onset, regularity, current velocity, water depth, spatial and temporal scale) (EOLSS, 2009).

## Effects of Heat on Comfort and Perceived Well-being

Temperature could affect human health both directly by heat stresses, which affect the body's ability to adjust and retain physiological function in response to the challenging thermal environment, and indirectly through pathways radiating from the detrimental effects of raised temperature on crop yield, socio-economic activity and disease transmission.

## Thermal comfort in an increasingly warm environment

Thermal comfort refers to an individual's appraisal of the thermal environment as being appropriate to his or her present wellbeing (ASHRAE, 1966). It is a socially determined sense of contentment with the environment emphasising neutral thermal sensations that do not impose demands on the individual's physiologic or adaptive resources (Hensen, 1991). Equanimity with the thermal environment occurs when the body temperature is maintained within a narrow range, sweating is limited to insensible loss and there is minimal physiological or behavioural adaptation effort (Djongyang *et al.*, 2010). The perception of environmental comfort is a cognitive process unique to each individual and is conditioned by physical, biological, psychological, cultural and lifestyle factors. Thermal sensations, therefore, differ between individuals in the same place at the same time. In addition, ratings of thermal comfort exhibit variation across different parts of the world due to specific comfort conditions associated with distinct regions of the world. These conditions are predominantly influenced by the climate in each geographic area (Fanger, 1973).

The comfort conditions for an individual are reproducible and will vary within a narrow limit from one day to the other and at different times of the day (Fanger, 1973). The variability of the comfort conditions during the day synchronises with the variability of the internal body temperature. The core body temperature is subject to the circadian rhythm: it tends to drop to the minimum during sleep hours and early in the morning, but increases during the day to a maximum in the late afternoon (Okamoto-Mizuno and Mizuno, 2012). Despite inter- and intra-personal variations in ratings of thermal comfort, given the same set of environmental conditions, it is traditionally thought that the optimal physiological conditions for comfort are present if the mean skin temperature is 33–34°C, with a concomitant minimised sweating rate and a low metabolic rate (Fanger, 1972). The sensation of thermal comfort is closely linked to skin temperature. When a person transits from a neutral thermal environment to a hot or cold microclimate, the changing perception of thermal load aligns with the skin temperature and sweat rate (Fanger, 1973).

### *Determinants of thermal comfort*

According to Macpherson (1962), the parameters that dictate thermal sensory perception are:

1. environmental factors: air temperature, air movement or velocity, mean radiant temperature and vapour pressure; and
2. human factors: activity level (metabolic rate) and clothing insulation.

A detailed quantitative assessment of these environmental and human factors enables effective characterisation of the thermal suitability of the environment. Specific combinations of these attributes modulate subjective assessments of thresholds for heat stress and the propensity for an adaptive response to thermal sensations. Other human factors that determine sensations of thermal satisfaction are age, gender, amount of subcutaneous fat, food and drink, and state of health (Auliciems and Szokolay, 2007). Furthermore, subjective assessments of comfort are contingent on the variability of these environmental factors and options available to the occupants of any environment to influence that variability (Yun, 2018). This is notably true for indoor spaces (Akande and Adebamowo, 2010).

## Elevated comfort temperatures

Global climate change has produced a trend towards increased frequency, magnitude and duration of extreme heat events. Urban areas are more severely impacted by the changing global climate because of the phenomenon of urban overheating – the urban heat island effect. It has been reported that more than 400 major cities of the world have experienced urban overheating in recent years (Santamouris, 2020).

Heatwaves further aggravate the urban heat island effect, degrading the thermal comfort of urban dwellers. The urban dwellers have to cope with a persistent trend of declining indoor comfort, particularly as the need for cooling becomes more pronounced during heatwaves and the rest of the summer period (Khan *et al.*, 2022).

Warmer regions of the world are more severely affected by temperature extremes than colder regions. Worsening comfort parameters and heat-related health outcomes are more consequential in the warmer regions (Baccini *et al.*, 2008). It is notable that the threshold temperature for health impact is higher in warmer regions of the world, and this would indicate some human adaptation to the relatively more strenuous thermal environment (Wei *et al.*, 2022). Overall, climate change conditions lead to an increase in cooling needs in both warmer and colder climates. At the same time, the heating needs in colder climates diminish as these regions experience warming (Khan *et al.*, 2022). A modelling study predicted for Toronto, the most populous city in Canada, an anticipated average decrease of 18–33% in heating energy demand and an average increase in cooling energy demand of 15–126% by the year 2070 (Berardi and Jafarpur, 2020).

A longer-term adaptive measure to address increasing environmental temperature is to improve the quality and appropriateness of housing. Residential and non-residential buildings have to be designed in such a way to maintain indoor thermoneutrality during the summer months and especially during extremes of temperature. During the warm summer months, indoor temperature can sometimes exceed outdoor temperature (Yun, 2018). As the indoor temperature increases, the comfort temperature threshold would also rise towards the upper range. Building characteristics that maintain the indoor comfort temperature even as the outdoor temperature increases should, therefore, be prioritised (Yun, 2018). Design options enabling passive cooling within the indoor space should be integrated with complementary measures to improve outdoor cooling in urban areas (this is further discussed in the section 'Adaptive Response to Excess Heat and Wetness Due to Climate Change').

## Humid heat and sleep insufficiency

Night-time sleep is facilitated by a drop in temperature. Conversely, warm nights impair sleep quality and the associated physiological restoration. Trends in night-time temperatures towards record high levels pose a substantial risk to sleep adequacy across the globe (Minor *et al.*, 2022). Real-world data has shown that the increase in night-time temperature across the entire observed range is harmful to sleep and depletes the physiological benefits sleep provides to the body (Minor *et al.*, 2022).

Sleep deprivation is a risk factor for cardiovascular disorder, failing cognitive function, mood disorders and fatiguability, depressed work and school performance, depressed immune response, metabolic disorder, suicide ideation and anger (Wang *et al.*, 2023). Sleep is therefore indispensable; it is a fundamental physiological function which complies with the circadian rhythm (Harding *et al.*, 2019). The circadian cycle is the biological daily modification of physical, mental and behavioural processes in response to the diurnal cycle of light and darkness (NGIMS, 2023). Mammals exhibit an adaptive thermoregulatory response to variations in environmental temperature, which prepares the body for sleep and maintains the sleep state. The body responds to heat by causing vasodilation in the hands and feet. This function reduces the core body and brain temperatures and induces sleepiness. Sleep onset corresponds to the circadian timing of the fastest drop in core body temperature, and the minimum core body temperature typically coincides with the deepest phase of sleep (Minor *et al.*, 2022).

Ideal room temperatures for sleep range between 19°C and 21°C. This corresponds with a thermoneutral skin microclimate of 31–35°C, at which the body can maintain its core temperature by regulating dry heat loss. An interference in the peripheral vasodilatory response can result when ambient temperature and humidity increase (Tempesta *et al.*, 2018). Sleep deficits tend to occur in these circumstances, and this can undermine the restorative and physiological benefits of sleep. The rise of temperatures beyond the thermoneutral zone leads to an increase in sleep latency and extended periods of wakefulness (Minor *et al.*, 2022). Humid heat negatively affects sleep indices such as sleep latency, wake after sleep onset, slow wave sleep, rapid eye movement sleep, sleep stages I (light) and III (deep), total sleep time and sleep efficiency index. As humid heat prevents efficient evaporative heat loss, there is a disruption of the physiologic regulation of core temperature and peripheral temperature necessary for sleep (Manzar *et al.*, 2012).

From the foregoing, it is unsurprising that night-time temperature-related sleep loss occurs more frequently in summer months than in winter. Similarly, temperature-related sleep loss tends to occur more frequently in warmer regions of the world than in colder regions. Thus, populations in the warmer parts of the world are at higher risk of sleep deprivation during temperature extremes and heatwaves. In the context of climate change, the projected susceptibility of these regions to temperature extremes underscores the difficulty the local populations will have in achieving restorative sleep (Roshan *et al.*, 2010).

Other than populations living in warmer regions, vulnerable groups also include older people (due to an age-related decline in thermoregulatory ability) and individuals in the lower socio-economic groups who may be unable to afford home appliances fitted to modify indoor temperature. People in less-resourced parts of the world are at higher risk of sleep loss from humid heat. In higher-income countries, the thermodynamic influence of humidity on sleep is more likely to be offset by climate-controlled indoor environments. However, in lower-income countries and among economically deprived populations, higher temperatures and humidity can attenuate restorative sleep due to the inability to afford climate-control equipment.

## The Urban Heat Island Effect

The urban heat island (UHI) describes the phenomenon in which built-up urban areas exhibit higher temperatures than their surrounding sub-urban and rural areas. Urban areas are more sensitive to elevated temperature because of the morphology of urban landscapes, higher heat emissions from human activities, and the relatively limited presence of nature-based cooling mechanisms for the urban environment. Urban areas may be warmer than the surrounding countryside by as much as 3.5–10°C (Voogt, 2002; Vardoulakis *et al.*, 2015). It is also well established that within cities, the suburban areas are cooler than the downtown precinct. The UHI is a consequence of a range of factors that collectively reduce heat dissipation in urban environments.

- Depletion of green and blue space: Urban areas are typically built up with lower vegetation cover in comparison with surrounding undeveloped countryside. In addition, many inland urban areas lack significant bodies of water. Evapotranspiration is an essential ecosystem service which facilitates environmental heat dissipation. Evapotranspiration is weaker in urban areas. There is also the loss of albedo provided by surface water.
- Architectural composition: The built-up morphology of cities undermines conditions for effective energy balance. The black asphalt, which is used to coat road surfaces, absorbs radiant energy from the sun during the day and releases the accumulated heat into the surrounding space at night. In a similar way, many materials used for building and construction absorb heat during the day and radiate the heat at night to the ambient environment. The street canyons, which are formed by adjoining tall buildings, impede the spatial dissipation of heat, trapping heat within the maze of towering architecture for longer periods.
- Waste heat: This is heat released from appliances, motor vehicles and machines that require energy for their operations. Examples include air conditioners, light bulbs, refrigerators and electrical fans, which generate and release heat as part of their operation.

The UHI intensifies the impacts of increasing heatwaves on cities in the context of climate change. Specifically, during summer and especially during heatwaves, the UHI effect has the potential to reinforce building overheating and precipitate the related health impacts (Heaviside *et al.*, 2017).

## Sustaining Health and Well-being in Temperature Extremes

### Thermal comfort and energy use

Electricity consumption varies substantially based on the temperature, climate and month of the year (Longden *et al.*, 2022). The thermal comfort standards of a geographic setting generally align with the energy consumption of surrounding buildings. Therefore, the thermal comfort compliance of a building is crucial to its sustainability (Yao *et al.*, 2009). Historical data indicate that urban overheating is responsible for a significant increase in energy demand for cooling. At sweltering temperatures, there is heavier dependence on air conditioning and higher electricity consumption. Consequently, peak electricity consumption during heat extremes will exceed patterns that are typical during periods of lower ambient temperature. Overheating trends historically align with energy consumption patterns. The cooling energy consumption trajectories for 18 cities were analysed for the period spanning 1970–2010. The cooling load for 1970 varied between 14 and 100 kWh/m$^2$/y, while the cooling load for 2010 was between 20 and 125 kWh/m$^2$/y. The increase in cooling load per year was on average 76 kWh/m$^2$/y/C across the cities (Santamouris, 2020).

A Taiwanese study assessed residential cooling and heating energy needs using the metrics *cooling degree days* (CDD) and *heating degree days* (HDD). The CDD represents the annual cumulative difference between the daily-averaged outdoor temperature and a reference temperature (comfort level), indicating the temperature whose exceedance signals that a building needs cooling. Similarly, for the HDD, the reference temperature would be the temperature below which the building needs to be heated. The investigators compared the projected incremental ratios of CDD for cities and towns in the 2025s to their values in the 1990s and found that, on average, urban overheating increased by 9.6% and 10.0%, respectively. The 2085s to 1990s incremental ratio revealed a 60% and 64.6% increase in cooling degree days for cities and towns, respectively. The projected HDDs for cities decreased by 15.1% while those for towns decreased by 13.3% in the 2025s. Similarly, the equivalent decrements in the 2085s compared to the 1990s were 73.3% and 68.4% degree days for cities and towns, respectively (Hwang *et al.*, 2017).

Hwang and colleagues (2017) also analysed the city-to-countryside and the town-to-countryside differences of the household-required cooling hours for three distinct periods: 1990s, 2025s and 2085s. In the 1990s, these differences were 1048 hours (80%) and 625 hours (48%), respectively. However, by the 2025s, the differences were 1080 hour (67%) and 634 hour (39%) for city-to-countryside and town-to-countryside, respectively. The increment was more modest in the 2085s, being 1107 hours (32%) and 624 hours (18%), respectively (Hwang *et al.*, 2017). The authors concluded that the propensity of the UHI effect to increase the relative urban-to-rural cooling hours diminished over time. This could be attributed to indoor temperatures in the countryside dwellings increasing enough to blunt the relative increments in cooling hours over time. Wei *et al.* (2022) conducted a study in 2019 to assess subjective and objective aspects of thermal comfort in Linshui, a rural area of China with a humid subtropical climate and average annual temperature of 20°C. The authors recorded an average indoor temperature of 29.24°C, which exceeded the annual whole-country (rural and urban) indoor average of 26.9°C. This discrepancy may be influenced by the quality of the rural housing and the extent to which it can maintain a comfortable indoor thermal environment.

### Temperature extremes and energy security justice

Rising temperatures adversely affect the demand and supply of electricity for indoor cooling. Peak electricity demand is often breached during the warmer months, synchronizing with increased cooling energy consumption (Santamouris, 2020). Temperature extremes may extend energy insecurity for lower-income households who cannot afford more resilient housing with efficient heating/cooling technology. The vulnerability of these households to extreme temperatures and the associated adverse health outcomes are magnified as a consequence (Walker and Day, 2012). Typically, these houses consume more electricity during heat extremes when the demand for cooling energy soars. Increasing electricity consumption signifies an imperative to expand existing power generation infrastructure to match the soaring demand. If additional power plants are constructed, the periods of peak demand will place enormous strain on the generation and transmission capacities of newer utility infrastructure. Moreover, additional power plants will increase GHG emissions, although some of the increased demand can be met using renewables, such as solar energy rooftop photovoltaic panels.

Longden *et al.* (2022) investigated the relationship between temperature extremes and the probability of electricity disconnection, and they observed that disconnections occurred more frequently during periods of extreme heat and cold, accounting for 29% of all disconnections. In addition, the study explored the probability of same-day disconnections on the first day of electricity subscription expiry. These disconnections were found to be more likely on days of temperature extremes and had a significantly higher probability of affecting households with higher consumption patterns. The authors noted that the households with the highest energy consumption patterns were among the most economically deprived. This is because these houses are generally of poorer quality and have fewer options for passive cooling. Therefore, existing energy insecurity will be amplified by climate change, compounding the health risks of vulnerable populations.

## Humid Heat and Disease

Rising temperatures and humidity can trigger or exacerbate disease outcomes in short- and longer-term exposure situations. Adverse health outcomes observed after shorter-term exposures are typically more acute, and transitory, but can be life-threatening. Chronic conditions can be exacerbated in single-exposure scenarios and in recurrent or chronic exposure situations.

### Primary Health Outcomes from Short-term Exposures

There are clinical conditions that occur more frequently during periods of temperature extremes and severe heat exposure scenarios. They are mostly acute but may have lasting complications. These conditions typically manifest as a consequence of fluid and electrolyte imbalances. They ensue during severe exposure episodes and progress along a continuum which indicates the increasing severity of the physiologic strain and the failure of normal adaptive mechanisms (Leon and Kenefick, 2010). They are collectively referred to as heat-related illnesses and are directly affected by heat stress.

Heat stress is the imposition of thermal load against which the body adjusts in order to maintain thermal balance. It results from a combination of internally generated heat coming from metabolic processes and extrinsic heat determined by factors such as prevailing radiant heat, ambient temperature, relative humidity, wind speed and clothing (Pinkerton and Rom, 2014). As heat stress approaches the human thermoregulatory limits, the physiological cost of responding to a thermal load that does not permit innate heat loss would bridle normal bodily function and predispose to disease. As the ability to downregulate the core body temperature declines, prolonged disturbances in body fluids and electrolytes composition would initiate clinical disease. The redirection of blood to the skin to facilitate evaporative heat loss reduces the blood volume; thus, there is diminished blood perfusion of the internal organs. The consequential drop in cardiac output results in tissue hypoxia (reduced oxygen supply) as organs are deprived of adequate blood supply. In more severe stress situations, cellular death and ischaemic tissue damage may occur, triggering an inflammatory response to injury, and inadvertently reinforcing the injury cascade (Leon and Kenefick, 2010; Pinkerton and Rom, 2014).

Heat strain is the cumulative response of the body to the physiologic challenge posed by heat stress (ACGIH, 2021). The adaptive capacity of the human body to elevated temperatures is limited by heat stress in times of temperature extremes and heatwaves (Sherwood and Huber, 2010). The clinical manifestations of heat stress may range from heat cramps, heat exhaustion, heat oedema and heat syncope to heatstroke, along an increasing spectrum of heat exposure. Heatstroke is characterised by core body temperatures of 40°C and above. This hyperthermic state has a high risk of fatality if treatment is not promptly administered. Severe heat stress, therefore, ranks among the leading causes of mortality from natural phenomena (Borden and Cutter, 2008).

In severe heat strain, hyperthermia, tissue ischaemia and oxidative stress resulting from reduced perfusion of tissues and organs can lead to tissue and organ damage affecting the heart, brain, kidneys, lungs and liver (Hanna and Tait, 2015). The brain may develop cerebral oedema due to cell injury and increased permeability of the blood–brain barrier. This permits toxic metabolites from the systemic blood circulation to infiltrate cerebral circulation. The toxic insult to the brain may manifest clinically as muscular weakness, cramps and outright neurological deficits (Beker *et al.*, 2018).

The heart rate increases in these hypovolaemic (low vascular blood volume) situations. However, hypotension (low blood pressure) develops due to dehydration and vasodilation resulting from the elaboration of nitric oxide, a pro-inflammatory mediator (Chaseling *et al.*, 2021). The heart tries to maintain the same output by overworking, but simultaneously, the heart is less nourished by its own diminished blood supply. This dysfunction in circulatory physiology would culminate in an acute cardiac failure or circulatory shock (Leon and Kenefick, 2010; Székely *et al.*, 2015). The respiratory system responds to raised core body temperature by increasing the volume of inhaled and exhaled air and accelerating the respiratory rate. This response facilitates respiratory heat loss through convection and evaporation, primarily cooling the brain (Beker *et al.*, 2018). Rapid oxygen–carbon dioxide exchange makes the blood more alkaline, a metabolic disorder known as respiratory alkalosis (Pinkerton and Rom, 2014). In addition, the lungs become inflammed and swollen, and the walls of lung arteries and capillaries lose structural integrity and begin to leak (Miyake, 2013; Pinkerton and Rom, 2014). As extreme heat events often occur concurrently with episodes of air pollution, the risk of respiratory insufficiency and secondary cardiac complications increases dramatically.

Heat-related depletion of body fluids and electrolyte imbalance can cause kidney cell death, acute renal injury (Chambers *et al.*, 2023), loss of the filtration capacity of the kidneys manifested as depressed urine volume (oliguria) and ultimately leads to acute renal failure (Kim *et al.*, 2018). The kidneys can also succumb to the toxic debris from damaged tissue which the blood transfers to the kidneys (Miyake, 2013). Similarly, the redirection of blood from the gastrointestinal system can cause symptoms such as nausea, vomiting, mucosal swelling and hepatic injury. Short-term severe strain on various organ systems not only increases morbidity, as evidenced by higher hospital admissions (Ogbomo *et al.*, 2017) and emergency department visits (Lin *et al.*, 2009; Bai *et al.*, 2016), but also increases the risk of mortality, as demonstrated by studies conducted in different parts of the world (Pan *et al.*, 1995; Tan *et al.*, 2010; Khatana *et al.*, 2022).

There is a rising incidence of heat stress due to global warming and the more frequent occurrence of heat extremes. Urban areas are more severely affected than rural areas (Wouters *et al.*, 2017). However, in the context of climate change, the number of heat stress days and nights has increased in both urban and rural areas (Oleson *et al.*, 2015). Similarly, tropical and subtropical climates are at higher risk of heat stress, with the tropics experiencing an increase in the number of heat stress nights (Fischer *et al.*, 2012). Nonetheless, peak heat stress, as measured by the wet-bulb temperature, displays remarkable similarity across diverse climatic zones (Sherwood and Huber, 2010).

## Health outcomes due to exacerbation of existing morbidities

Pre-existing cardiovascular, respiratory, renal and metabolic diseases can be exacerbated or aggravated (acute-on-chronic episodes) by the redistribution of fluid from the organs and tissues to the skin (Chambers *et al.*, 2023). Increased cardiovascular efforts to compensate for the reduced blood volume may result in distortions of the cardiac rhythm and myocardial infarction due to diminished blood supply to heart muscles and tissues (Chaseling *et al.*, 2021). Consequently, the risks of cardiac ischaemia, myocardial infarction and cardiac failure significantly increase.

Injury to the renal tubular cells stimulates healing by repair rather than regeneration. Repeated cycles of tissue death and repair would result in fibrosis and loss of the functional ability of the kidney. The steady-state clinical features of chronic respiratory diseases, such as bronchial asthma, chronic obstructive pulmonary disease and bronchiectasis, are exacerbated by heat stress. Moreover, the frequent co-occurrence of heat stress and air pollution episodes consolidates the exposure burden to individuals with chronic respiratory diseases. The inflammatory response of the body to heat stress also involves the respiratory organs; thus, acute inflammation of the bronchi, bronchioles and alveoli can superimpose on existing respiratory disorders.

Diabetes mellitus is another disease that is sensitive to extreme heat events. In diabetes, inadequate thermoregulation occurs due to reduced blood flow to the skin and sweating. There is also an impairment of the autonomic nervous systemic responses. Diabetes predisposes individuals to fluid and electrolyte imbalance, which could be compounded by sweat loss during extreme heat events. With reduced blood volume, haemoconcentration occurs, making the blood more coagulable. Diabetes-related complications such as chronic kidney disease and ischaemic heart disease tend to occur more often during extreme heat. In addition, therapeutic preparations for diabetes management such as insulin are heat sensitive and could easily denature in extreme heat (Vallianou *et al.*, 2021).

## Consequences of Intense Precipitation and Floods

Flooding happens when the water level in an existing body of water rises beyond its natural banks or constructed embankment and submerges predominantly dry places. Flooding also occurs when floodplains and low-lying areas become submerged by runoff water after heavy rains. The consequences of floods can be grave, with far-reaching ramifications for the socio-economic life and wellbeing of people in the affected communities. These consequences include, damage to property; bodily harm and disease; disruption of essential municipal services; and the destruction of livelihoods (Nicholls *et al.*, 2015).

- Damage and destruction of property: The ability of a building to withstand a flood depends on its location, age, structural integrity and specification (Nikolowski *et al.*, 2013). The walls of a building can suffer moisture damage after heavy rains and floods, but this is a more gradual process. On the other hand, the hydrostatic pressure of water build-up against the walls can cause sudden cracks or collapse (Nikolowski *et al.*, 2013). The indoor seepage of water soaks the living space, furniture, stored foods and personal items. The personal economic loss can be considerable as the affected items may be irrecoverable. Flood-impacted buildings may become unsafe for further human dwelling, prompting a need for relocation.
- Bodily harm and disease: Individuals may suffer injuries from physical trauma caused by falling objects or objects swept in the tow of fast-moving water currents. Physical trauma can be a cause of immediate death. Hypothermia can be an immediate threat from prolonged contact with surrounding cold water. Electrocution is a risk when power lines and exposed electricity sources and conductors are in contact with floodwaters (Vardoulakis *et al.*, 2015). Drowning is a common cause of death in these situations. Many drownings occur in vehicles that are swept into deeper floodwater. Drowning in the residential dwelling mostly affects the elderly and people with physical or cognitive disability (Ahern *et al.*, 2005). In the aftermath of flood events, feacal-oral transmission of disease can occur in instances where sanitary utilities have been damaged and drinking water sources have been contaminated. The incidence of vector-borne diseases can may increase because of stagnant pools of water which provide breeding sites for

mosquitoes. Rodent-borne diseases such as Leptospirosis can also breakout as a result of the displacement of vermin from their habitats. Mental health problems such as post-traumatic stress disorder, anxiety, depression, sleeplessness and irritability are serious consequences of floods on human populations (Ahern *et al.*, 2005). Impaired mental health can prevent recovery and restoration to normal productive life.

- Interruption of municipal services and amenities: Floods can interrupt crucial municipal services such as healthcare and emergency services, and thereby indirectly impact health by degrading the healthcare support system. The damage or destruction of public waterworks and sewage systems may release biohazards and create conditions amenable to infectious disease outbreaks and epidemics (Nicholls, 2018). The inability of emergency and rescue services to access the locations of flood victims undermines public health responses in these circumstances.
- Destruction of livelihoods: Floods can lead to the destruction of farmlands, grazing land and livestock, irrigation systems and aquaculture, as well as commercial and industrial infrastructure. The community may lose its economical viability and competitiveness (Nicholls *et al.*, 2015). Social systems are broken and local populations are dispossessed and reliant on relief aid to survive until the viability of society can be restored.

## Adaptive Response to Excess Heat and Wetness Due to Climate Change

### Heat adaptation

A multi-faceted response is required to adapt to an increasingly hostile thermal environment. The response options should preserve human health without exacerbating GHG emissions or unsustainable resource use. The adaptive response must aim to reduce exposure and improve health resilience through feasible and sustainable long-term strategies and measures. An appropriate response requires competences that are specific to the health sector deployed in tandem with competences from other sectors. An oversight agency has to be mandated by the government to coordinate input and participation from diverse sectors, with a strong political commitment and a sense of urgency in implementing recommended solutions. These solutions should comply with global best practices, but can be tailored to reflect the peculiarities of local settings (Corburn, 2009). An adaptive response should incorporate elements of the strategies discussed below. These strategies should be developed and implemented as part of a local, regional or national heat-health rolling response plan.

#### *Stakeholder considerations and enrolment*

A comprehensive list should be compiled of stakeholders responsible for planning and executing response strategies as well as individuals or groups who will be crucial in securing a citizen buy-in and participation. Stakeholders should include: ministries of health, local and regional health authorities, town planners, civil engineers, meteorologists, public works departments, research organisations, media organisations, civil society groups and community representatives. The composition of the stakeholders would align with the level (tactical, strategic or operational) at which planning and execution occur.

Political authority, technical competence, understanding of contextual realities and the local resource inventory have to be harnessed within a partnership framework that encourages broad participation while being coordinated by a designated authority or agency. Ideally, globally acknowledged state-of-the-art practices must interface with local capacities to co-produce and co-design solutions (Corburn, 2009).

#### *Integrated solutions*

A multi-pronged heat adaptation strategy should involve a combination of sustainable methods to reduce heat exposure. These should include short-term interventions, such as heat-health alerts as a response to a heatwave, and long-term interventions aiming to reduce urban overheating by modifying our environment, for example by increasing green space and shading. Some of these solutions (suggested below) may require significant investment at the outset, but they are easier to maintain and cost-beneficial in the long term.

**INFRASTRUCTURAL PLANNING AND DESIGN.**

- This involves nature-based solutions such as parks and green spaces. These spaces reduce the local temperature by increasing shading and evapotranspiration. The effectiveness of these benefits could be enhanced if parks and green spaces feature more tree cover rather than just grass and decorative plants. Trees offer better shade and provide a visual and noise buffer to local traffic and other pollution sources. Well-designed green spaces also contribute to reducing air pollution (which frequently occurs in tandem with heat extremes), creating shade and ambient cooling, collectively mitigating the damaging effect of heat on human health. In addition to providing a cooler microclimate, green spaces also provide recreational benefits to local populations. The green space design should match the surrounding urban area and the local climate, enhancing the overall benefits (Coffel *et al.*, 2018). Furthermore, the design of the green space, choice of trees and other vegetation species, and quality of their canopy should be resilient to the increasingly harsh thermal environment (Lin *et al.*, 2021). Unintended negative effects related to allergenic pollen and biogenic volatile organic compounds emitted from certain plant species, as well as water and energy requirements for maintenance, should be carefully considered in the design of green spaces (Salmond *et al.*, 2016).
- Passive cooling solutions involve methods and designs aimed at maintaining appropriate indoor comfort conditions with minimum energy consumption (Bhamare *et al.*, 2019). This can be achieved through a combination of methods that protect the building envelope from direct heat, modulate the heat gain by the building and facilitate the dissipation of heat from the building into environmental heat sinks (Bhamare *et al.*, 2019). In convective cooling, natural ventilation enables the air to act as a sink that collects excess heat from a building and transmits it to the atmosphere. This is achieved through wind-driven ventilation and buoyancy (Oropeza-Perez and Østergaard, 2018). Wind-driven ventilation functions by the principle of a positive pressure differential between the windward facade of the building and the leeward side of the building. The static pressure on the building envelope depends on climatic factors such as the air density, mean pressure field at ventilation openings, wind direction, wind speed and incident angle of wind streams, and building factors such as the plan area density, facade orientation and frontal aspect ratio (Oropeza-Perez and Østergaard, 2018). Passive cooling can also be achieved by increasing the thermal mass of the building; thus the structural material of the building provides thermal resistance by absorbing excess heat through radiation and convection, thereby preventing the transfer of outdoor heat to indoor (Givoni, 2011). The structural material can absorb either the sensible heat or the latent heat, which it radiates back to the environment at night. The effectiveness of the thermal mass in achieving passive cooling depends on local climatic conditions, the building orientation, construction and material properties (Bhamare *et al.*, 2019).
- Solar shades control the amount of incident solar radiation on a building envelope, which translates to an eventual indoor heat gain. Natural shading can be provided by trees and vegetation, acting as shields for roofs, walls and windows. The shielding canopy and evapotranspiration by plants can lower area temperature by up to 5°C. Further, shading can also be achieved by installing horizontal louvers, overhangs, awnings, light shelves and blind system (Kamal, 2012). Cool roofs (i.e. high reflectance roofs that absorb less heat from the sun) can also increase the albedo of the urban canopy and reduce building overheating (Macintyre and Heaviside, 2019). A range of construction materials and

engineering solutions can reduce urban heat islands in cities (Santamouris, 2020).

- Urban morphology and building designs: the city size, spatial layout, geometry of streets and buildings, street width and the thermal properties of building facades contribute to the urban heat island effect (Zou and Zhang, 2021). Tall buildings on either side of streets retain air pollutants and heat. This relationship is more prominent with greater building height-to-street-width ratio. These street canyons decrease radiative cooling, as typically, the surrounding buildings absorb the outgoing long-wave radiation, hindering their escape into the atmosphere (Coffel *et al.*, 2018). Reconceptualising urban planning so that urban centres are characterised by more pedestrian walks, green spaces, green building walls, green roofs, reflective tarmacs and roofs, flowing urban water bodies and fountains can substantially lower urban local near-surface temperatures (Coffel *et al.*, 2018).

### *Health, social and municipal services*

A crucial part of a heat-health action plan is the establishment of early warning systems. These systems are designed to forestall the harmful health impacts of extreme heat events (NDMAGI, 2019). They integrate historical meteorological and health surveillance data, as well as weather forecasts, to predict the onset and duration of periods of extreme heat and humidity that are likely to result in severe health outcomes and increased mortality (Jay *et al.*, 2021). Thresholds for heat advisories are typically predicated upon local exposure–response relationships (Ebi, 2019). Early warning systems do not only issue heat alerts but also provide instructions on self-protective measures. The warnings are more effective when targeting specific high-risk and vulnerable groups (Coffel *et al.*, 2018).

The early warning system can be part of an overall information strategy which raises the public awareness of the risks associated with periods of extended heat and provides advice on behavioural changes for improved health and comfort during the warm summer months – for example, wearing appropriate clothing, staying hydrated, reducing strenuous physical activity during the hotter periods of the day and avoiding direct sunlight exposure (Matthies *et al.*, 2008). Indigenous and local knowledges on adapting to changing environmental conditions can also provide valuable insights into community resilience to extreme heat (Standen *et al.*, 2022). The communication strategy should be adapted to the specific needs of diverse population and occupational groups, including the elderly, homeless people, culturally and linguistically diverse communities, outdoor workers in construction, miners, athletes and workers employed in service sectors that provide health and social support services.

Healthcare providers need tailored training to promptly diagnose heat-related illnesses and to initiate timely treatment. Providing information on ways to prevent heat strain should become a routine healthcare service. Medical practitioners should also identify high-risk patients, such as those with cardiovascular diseases, people living with asthma or diabetes, and people on specific medications such as diuretics, antidepressants, anticholinergic drugs, and illegal narcotics which tend to affect fluid–electrolyte balance or the ability to sweat. High-risk patients should be booked for follow-up visits. In addition, home visits would allow for an assessment of the physical attributes of the domestic environment, so that tailored advice could be given to optimise dwellings against heat stress susceptibility (Matthies *et al.*, 2008). It is also important that social services identify residents at risk (e.g. elderly people living alone) and ensure that they receive adequate health advice and support during heatwaves.

Another municipal service is the establishment of cooling centres or heat refuges. These are naturally cooled or air-conditioned buildings open to the public for extended hours (Dufty, 2022). Cooling centres may be public buildings such as community centres, libraries, religious centres or schools that can provide safety during heat extremes. These centres are particularly helpful for individuals who cannot afford the cost of domestic air conditioning or other modifications to their home environment. Cooling centres are often integrated with early warning systems, in which case, the address and phone numbers of the cooling centres are provided as part of the heat alert (Widerynski *et al.*, 2017). Cooling centres are often under the administration of local governments, but the majority of their operations are often facilitated by voluntary workers. Cooling centres are limited because they provide sanctuary from heat only in daytime hours. With rising night-time temperatures, these vulnerable populations may still be at risk from heat during night-time hours spent at home.

### *Monitoring and evaluation*

An effective heat-health action plan should include an evaluation component that can generate evidence to demonstrate the effectiveness and efficiency of the plan (Dwyer *et al.*, 2022). High-quality evaluation is an essential tool for decision making, as it addresses the heat-protection needs of the most vulnerable people in society, thereby mitigating health inequities (Ebi, 2019). The evaluation framework should be robust and incorporate instruments for assessing both morbidity and mortality, including the distributional effects of heat adaptation (Dwyer *et al.*, 2022).

Moreover, the framework should comprise both short-term and long-term indicators of effectiveness and efficiency. It could also consider alternative explanations for the observed intervention outcomes, either measured directly or by using proxy indices. A heat-health plan evaluation should conform to the basic schedule of a public health intervention evaluation plan. This involves incorporating process evaluation to assess if implementation of the intervention complied with proposed standards, and outcome evaluation to assess the efficacy of morbidity and mortality reduction, and improvements in community resilience and health equity (Matthies *et al.*, 2008).

## Flood Management and Resilience

Resilience is a concept which is triple-pronged: (i) ecological resilience denotes the ability of a site to adapt to extraneous shocks by modifying and restraining its response to these shocks; (ii) engineering resilience signifies the potential of the site to re-enact its pre-disturbance state,; (iii) community resilience is the ability of a community to respond, adapt to, and recover from adverse events (Pickett *et al.*, 2014; Vardoulakis *et al.*, 2022). Flood resilience minimises damage from floods while lessening recovery time and effort (Nicholls *et al.*, 2015). It requires a combination of institutional, technological and managerial instruments which include some of the strategies that have been discussed above, such as broad stakeholder engagement and participation, infrastructural redesign and early warning systems (Vardoulakis *et al.*, 2022). Flood resilience supplements traditional flood risk management with approaches that are pertinent to climate-related disaster risk reduction.

Traditional flood management systems entail the construction of physical barriers like dams and embankments; planting flood-resistant vegetation which would absorb water and mitigate the momentum of runoffs; cutting terraces into hillsides to reduce the velocity of waterflow; and construction of drainage channels and ponds. It also encompasses flood-resistant housing, food-stock preparedness and disaster relief (Itsukushima *et al.*, 2021).

Flood resilience targets and blunts vulnerabilities that magnify the impacts of floods. It might combine traditional flood management methods with land-use planning, flood-risk estimation, early warning systems, evacuation and emergency plans or provision of an insurance policy (Nicholls *et al.*, 2015). Components of a flood-resilient architecture at the regional level may include but are not limited to:

- An infrastructural framework that includes stormwater drainage, levees, dykes, embankments, dams and reservoirs. Controlled release from reservoirs at pre-determined times can reduce the scale of floods when they occur.
- Emergency financing to assist individuals and families in regaining their socio-economic viability.
- Community-based planning that takes on board the local risk situation and lessons from previous floods or floods that occurred elsewhere.
- An appropriate policy and regulatory framework that caters to the risk profiles, vulnerabilities and needs of individuals,

families, groups and organisations. Broad-based inclusion of different segments of society will encourage participation, ownership and strengthening of the policy process.

- Incorporating climate change scenarios into economic modelling and the development of projections to stay ahead of evolving flood-risk scenarios.

At the individual and family levels, resilience may be improved by building on high ground and adaptation of the building stock to prevent the ingress of water during flood scenarios. Buildings should be made wet-proof using construction materials that make damage-prone walls, ceilings and roofs significantly resistant to flood damage, or by using constructions materials that are less resistant to damage but are cheap and can be quickly replaced (Nikolowski *et al.*, 2013). Buildings should be covered by sturdy well-anchored roofs that are unlikely to be blown off during storms. Electrical connections and meters should be raised beyond heights where floodwaters can reach them.

## Conclusion

Dealing with the environmental health impacts of a hotter and damper climate requires strong collaboration across multiple sectors and policy portfolios, including environment agencies, health and emergency services and urban planning departments. It should be stressed that community engagement, involving people who are at higher risk because of their age, health condition, socio-economic situation or cultural and linguistic background, is a crucial factor for the successful design and implementation of heat-health and flood management and community resilience plans, and other climate adaptation strategies. Drawing from diverse knowledge systems, including Indigenous and Western science and data, can generate innovative solutions to environmental change and weather extremes. In addition to climate-proofing infrastructure and services, building community cohesion and social capital will strengthen our resilience to a hotter and damper climate and to more variable weather conditions under climate change.

## References


Abram, N.J., Henley, B.J., Sen Gupta, A., Lippmann, T.J.R., Clarke, H. *et al.* (2021) Connections of climate change and variability to large and extreme forest fires in Southeast Australia. *Communications Earth & Environment* 2, 8.

ACGIH (2021) *TLVs and BEIs Based on the Threshold Limit Values and Biological Exposure Indices for Chemical Substances and Physical Agents*. Signature Publications, Cincinnati, Ohio.

Ahern, M., Kovats, R.S., Wilkinson, P., Few, R. and Matthies, F. (2005) Global health impacts of floods: Epidemiologic evidence. *Epidemiologic Reviews* 27, 36–46.

Akande, O. and Adebamowo, M. (2010) Indoor thermal comfort for residential buildings in hot-dry climate of Nigeria. Proceedings of conference – Adapting to Change: New Thinking on Comfort Cumberland Lodge, pp. 9–11.

ASHRAE Standard (1966) *Thermal Comfort Conditions for Human Occupancy*. American Society of Heating, Refrigerating and Air-Conditioning Engineers Standards Engineers, Atlanta, pp. 55–66.

Auliciems, A. and Szokolay, S.V. (2007) *Thermal Comfort*. PLEA Notes: Design Tools and Techniques 3. Passive and Low Energy Architecture International Brisbane, Australia. Available at: https://dev.humanitarianlibrary.org/sites/default/files/2014/02/plea_2007_thermal_comfort.pdf (accessed 18 July 2025).

Baccini, M., Biggeri, A., Accetta, G., Kosatsky, T., Katsouyanni, K. *et al.* (2008) Heat effects on mortality in 15 European cities. *Epidemiology* 19, 711–719.

Bai, L., Li, Q., Wang, J., Lavigne, E., Gasparrini, A. *et al.* (2016) Hospitalizations from hypertensive diseases, diabetes, and arrhythmia in relation to low and high temperatures: Population-based study. *Scientific Reports* 6, 30283.

Barreca, A.I. (2012) Climate change, humidity, and mortality in the United States. *Journal of Environmental Economics and Management* 63, 19–34.

Beker, B., Cervellera, C., Vito, A. and Musso, C. (2018) Human physiology in extreme heat and cold. *International Archives of Clinical Physiology* 1, 1–8.

Berardi, U. and Jafarpur, P. (2020) Assessing the impact of climate change on building heating and cooling energy demand in Canada. *Renewable and Sustainable Energy Reviews* 121, 109681.

Bhamare, D.K., Rathod, M.K. and Banerjee, J. (2019) Passive cooling techniques for building and their applicability in different climatic zones—the state of art. *Energy & Buildings* 198, 467–490.

Borden, K.A. and Cutter, S.L. (2008) Spatial patterns of natural hazards mortality in the United States. *International Journal of Health Geographics* 7, 64.

Budd, G.M. (2008) Wet-bulb globe temperature (Wbgt)–its history and its limitations. *Journal of Science and Medicine in Sport* 11, 20–32.

Chambers, E.C., Norris, K.C., Levano, S.R. and Golestaneh, L. (2023) Extreme heat can exacerbate disproportionate burden of severe kidney disease in historically marginalized communities: Call to action. *Clinical Journal of the American Society of Nephrology* 18(8), 1089–1091.

Chaseling, G.K., Iglesies-Grau, J., Juneau, M., Nigam, A., Kaiser, D. *et al.* (2021) Extreme heat and cardiovascular health: What a cardiovascular health professional should know. *Canadian Journal of Cardiology* 37, 1828–1836.

Christidis, N., Jones, G.S. and Stott, P.A. (2015) Dramatically increasing chance of extremely hot summers since the 2003 European heatwave. *Nature Climate Change* 5, 46–50.

Coffel, E.D., De Sherbinin, A., Horton, R.M., Lane, K., Kienberger, S. *et al.* (2018) The science of adaptation to extreme heat. In: Zommers, Z. and Alverson, K. (eds) *Resilience.* Elsevier, Amsterdam, pp. 89–103.

Corburn, J. (2009) Cities, climate change and urban heat island mitigation: Localising global environmental science. *Urban Studies* 46, 413–427.

Coumou, D., Robinson, A. and Rahmstorf, S. (2013) Global increase in record-breaking monthly-mean temperatures. *Climatic Change* 118, 3–4.

Cox, D.T.C., Maclean, I.M.D., Gardner, A.S. and Gaston, K.J. (2020) Global variation in diurnal asymmetry in temperature, cloud cover, specific humidity and precipitation and its association with leaf area index. *Global Change Biology* 26, 7099–7111.

Davis, R.E., Mcgregor, G.R. and Enfield, K.B. (2016) Humidity: A review and primer on atmospheric moisture and human health. *Environmental Research* 144, 106–116.

Djongyang, N., Tchinda, R. and Njomo, D. (2010) Thermal comfort: A review paper. *Renewable and Sustainable Energy Reviews* 14, 2626–2640.

Dufty, N. (2022) Using heat refuges in heatwave emergencies. *The Australian Journal of Emergency Management* 37, 38–44.

Dwyer, I.J., Barry, S.J.E., Megiddo, I. and White, C.J. (2022) Evaluations of heat action plans for reducing the health impacts of extreme heat: Methodological developments (2012–2021) and remaining challenges. *International Journal of Biometeorology* 66, 1915–1927.

Ebi, K.L. (2019) Effective heat action plans: Research to interventions. *Environmental Research Letters* 14, 122001.

Ebi, K.L., Capon, A., Berry, P., Broderick, C., De Dear, R. *et al.* (2021) Hot weather and heat extremes: Health risks. *The Lancet* 398, 698–708.

EOLSS (2009) *Water Interactions with Energy, Environment, Food, and Agriculture.* UNESCO-EOLSS, Oxford, UK.

Fanger, P.O. (1972) *Thermal Comfort: Analysis and Applications in Environmental Engineering.* Mcgraw-Hill, New York.

Fanger, P.O. (1973) Assessment of man's thermal comfort in practice. *British Journal of Industrial Medicine* 30, 313.

Fischer, E. and Knutti, R. (2015) Anthropogenic contribution to global occurrence of heavy-precipitation and high-temperature extremes. *Nature Climate Change* 5(6).

Fischer, E.M., Oleson, K.W. and Lawrence, D.M. (2012) Contrasting urban and rural heat stress responses to climate change. *Geophysical Research Letters* 39(3).

Givoni, B. (2011) Indoor temperature reduction by passive cooling systems. *Solar Energy* 85, 1692–1726.

Guo, Y., Gasparrini, A., Armstrong, B.G., Tawatsupa, B., Tobias, A. *et al.* (2017) Heat wave and mortality: A multicountry, multicommunity study. *Environmental Health Perspectives* 125, 087006.

Hanna, E.G. and Tait, P.W. (2015) Limitations to thermoregulation and acclimatization challenge human adaptation to global warming. *International Journal of Environmental Research and Public Health* 12, 8034–8074.

Harding, E.C., Franks, N.P. and Wisden, W. (2019) The temperature dependence of sleep. *Frontiers in Neuroscience* 13, 336.

Havenith, G. (2003) Interaction of clothing and thermoregulation. *Exogenous Dermatology* 1, 221–230.

Heaviside, C., Macintyre, H. and Vardoulakis, S. (2017) The urban heat island: Implications for health in a changing environment. *Current Environmental Health Reports* 4, 296–305.

Hensen, J. (1991) On the thermal interaction of building structure and heating and ventilating system. PhD thesis, Eindhoven University of Technology, Eindhoven, Netherlands.

Horton, R.M., Mankin, J.S., Lesk, C., Coffel, E. and Raymond, C. (2016) A review of recent advances in research on extreme heat events. *Current Climate Change Reports* 2, 242–259.

Hwang, R.L., Lin, C.Y. and Huang, K.T. (2017) Spatial and temporal analysis of urban heat island and global warming on residential thermal comfort and cooling energy in Taiwan. *Energy & Buildings* 152, 804–812.

IPCC (2023) *Synthesis Report. Contribution of Working Groups I, II and III to the Sixth Assessment Report of the Intergovernmental Panel on Climate Change*. IPCC, Geneva, Switzerland.

Itsukushima, R., Ohtsuki, K. and Sato, T. (2021) Learning from the past: Common sense, traditional wisdom, and technology for flood risk reduction developed in Japan. *Regional Environmental Change* 21(3).

Jay, O., Capon, A., Berry, P., Broderick, C., De Dear, R. *et al.* (2021) Reducing the health effects of hot weather and heat extremes: From personal cooling strategies to green cities. *The Lancet* 398, 709–724.

Kamal, M.A. (2012) An overview of passive cooling techniques in buildings: Design concepts and architectural interventions. *Acta Technica Napocensis: Civil Engineering & Architecture* 55, 84–97.

Khan, S.H., Paolini, R., Caccetta, P. and Santamouris, M. (2022) On the combined impact of local, regional, and global climatic changes on the urban energy performance and indoor thermal comfort—the energy potential of adaptation measures. *Energy & Buildings* 267, 112152.

Khatana, S.A.M., Werner, R.M. and Groeneveld, P.W. (2022) Association of extreme heat and cardiovascular mortality in the United States: A county-level longitudinal analysis from 2008 to 2017. *Circulation* 146, 249–261.

Kim, E., Kim, H., Kim, Y.C. and Lee, J.P. (2018) Association between extreme temperature and kidney disease in South Korea, 2003–2013: Stratified by sex and age groups. *Science of the Total Environment* 642, 800–808.

Leon, L.R. and Kenefick, R. (2010) Pathophysiology of heat-related illnesses. In: Auerbach, P.S. (ed.) *Wilderness Medicine Textbook*, 6th edn. Mosby, Philadelphia, pp. 215–231.

Lin, B.B., Ossola, A., Alberti, M., Andersson, E., Bai, X. *et al.* (2021) Integrating solutions to adapt cities for climate change. *The Lancet Planetary Health* 5, E479–E486.

Lin, S., Luo, M., Walker, R.J., Liu, X., Hwang, S.A. *et al.* (2009) Extreme high temperatures and hospital admissions for respiratory and cardiovascular diseases. *Epidemiology* 20, 738–746.

Longden, T., Quilty, S., Riley, B., White, L.V., Klerck, M. *et al.* (2022) Energy insecurity during temperature extremes in remote Australia. *Nature Energy* 7, 43–54.

Macintyre, H.L. and Heaviside, C. (2019) Potential benefits of cool roofs in reducing heat-related mortality during heatwaves in a European city. *Environment International* 127, 430–441.

Macpherson, R.K. (1962) The assesment of the thermal environments. A review. *British Journal of Industrial Medicine* 19(3), 151–164.

Manzar, M.D., Sethi, M. and Hussain, M.E. (2012) Humidity and sleep: A review on thermal aspect. *Biological Rhythm Research* 43, 439–457.

Masterton, J.M. and Richardson, F.A. (1979) *Humidex: A Method of Quantifying Human Discomfort Due to Excessive Heat and Humidity*. Environment Canada, Ontario, Canada. Available at: https://publications.gc.ca/site/fra/9.865813/publication.html (accessed 18 July 2025).

Matthews, T. (2018) Humid heat and climate change. *Progress in Physical Geography: Earth and Environment* 42, 391–405.

Matthies, F., Bickler, G., Marín, N.C. and Hales, S. (2008) *Heat-Health Action Plans*. WHO Regional Office for Europe, Copenhagen.

Minor, K., Bjerre-Nielsen, A., Jonasdottir, S.S., Lehmann, S. and Obradovich, N. (2022) Rising temperatures Erode human sleep globally. *One Earth* 5, 534–549.

Miyake, Y. (2013) Pathophysiology of heat illness: Thermoregulation, risk factors, and indicators of aggravation. *Japan Medical Association Journal* 56(3), 167–173.

NDMAGI (2019) *National Guidelines for Preparation of Action Plan – Prevention and Management of Heat Wave*. National Disaster Management Authority, Ministry of Home Affairs, New Delhi.

NGIMS (2023) What are Circadian Rhythms? National Institute of General Medical Sciences. Available at: https://Www.Nigms.Nih.Gov/Education/Fact-Sheets/Pages/Circadian-Rhythms.Aspx#:~:Text=Circadian%20rhythms%20are%20physical%2c%20mental,The%20study%20of%20circadian%20rhythms (accessed 2 November 2023).

Nicholls, R., Zanuttigh, B., Vanderlinden, J.P., Weisse, R., Silva, R. *et al*. (2015) Developing a holistic approach to assessing and managing coastal flood risk. In: Zanuttigh, B., Nicholls, R., Vanderlinden, J.P., Burcharth, H.F. and Thompson, R.C. (eds) *Coastal Risk Management in a Changing Climate*. Butterworth-Heinemann, Boston, Massachusetts, pp. 9–53.

Nicholls, R.J. (2018) Adapting to sea-level rise. In: Zommers, Z. and Alverson, K. (eds) *Resilience: The Science of Adaptation to Climate Change*. Elsevier, Amsterdam, pp. 13–29.

Nikolowski, J., Goldberg, V., Zimm, J. and Naumann, T. (2013) Analysing the vulnerability of buildings to climate change: Summer heat and flooding. *Meteorologische Zeitschrift* 22, 145–153.

Ogbomo, A.S., Gronlund, C.J., O'Neill, M.S., Konen, T., Cameron, L. *et al*. (2017) Vulnerability to extreme-heat-associated hospitalization in three counties in three counties in Michigan, USA, 2000–2009. *International Journal of Biometeorology* 61, 833–843.

Okamoto-Mizuno, K. and Mizuno, K. (2012) Effects of thermal environment on sleep and circadian rhythm. *Journal of Physiological Anthropology* 31, 14.

Oleson, K.W., Monaghan, A., Wilhelmi, O., Barlage, M., Brunsell, N. *et al*. (2015) Interactions between urbanization, heat stress, and climate change. *Climatic Change* 129, 525–541.

Oropeza-Perez, I. and Østergaard, P.A. (2018) Active and passive cooling methods for dwellings: A review. *Renewable and Sustainable Energy Reviews* 82, 531–544.

Pan, W.H., Li, L.A. and Tsai, M.J. (1995) Temperature extremes and mortality from coronary heart disease and cerebral infarction in elderly Chinese. *The Lancet* 345, 353–355.

Pickett, S.T.A., Mcgrath, B., Cadenasso, M. and Felson, A. (2014) Ecological resilience and resilient cities. *Building Research and Information* 42(2).

Pinkerton, K.E. and Rom, W.N. (2014) *Global Climate Change and Public Health*. Humana Press, New York.

Pizzorni, M., Innocenti, A. and Tollin, N. (2024) Droughts and floods in a changing climate and implications for multi-hazard urban planning: A review. *City and Environment Interactions* 24, 100169.

Rachid, A. and Qureshi, A.M. (2023) Sensitivity analysis of heat stress indices. *Climate* 11(9), 181.

Raymond, C., Matthews, T. and Horton, R.M. (2020) The emergence of heat and humidity too severe for human tolerance. *Science Advances* 6, Eaaw1838.

Robinson, P.J. (2000) Temporal trends in United States dew point temperatures. *International Journal of Climatology* 20, 985–1002.

Roshan, G.R., Ranjbar, F. and Orosa, J.A. (2010) Simulation of global warming effect on outdoor thermal comfort conditions. *International Journal of Environmental Science & Technology* 7, 571–580.

Salmond, J.A., Tadaki, M., Vardoulakis, S., Arbuthnott, K., Coutts, A. *et al*. (2016) Health and climate related ecosystem services provided by street trees in the urban environment. *Environmental Health* 15, S36.

Santamouris, M. (2020) Recent progress on urban overheating and heat island research. Integrated assessment of the energy, environmental, vulnerability and health impact. Synergies with the global climate change. *Energy & Buildings* 207, 109482.

Seneviratne, S.I., Donat, M.G., Mueller, B. and Alexander, L.V. (2014) No pause in the increase of hot temperature extremes. *Nature Climate Change* 4, 161–163.

Sherwood, S.C. and Huber, M. (2010) An adaptability limit to climate change due to heat stress. In: *Proceedings of the National Academy of Sciences*, pp. 9552–9555 (Vol. 107).

Solomon, S., Qin, D., Manning, M., Chen, Z., Marquis, M. *et al*. (2007) *Climate Change 2007: The Physical Science Basis*. Working Group I Contribution to the Fourth Assessment Report of the IPCC, IPCC, Geneva, Switzerland.

Standen, J.C., Spencer, J., Lee, G.W., Van Buskirk, J., Matthews, V. *et al*. (2022) Aboriginal population and climate change in Australia: Implications for health and adaptation planning. *International Journal of Environmental Research and Public Health* 19(12), 7502.

Steadman, R. (1994) Norms of apparent temperature in Australia. *Australian Meteorological Magazine* 43, 1–16.

Steffen W, Persson A, Deutsch L, Zalasiewicz J, Williams M, Richardson K, Crumley C, Crutzen P, Folke C, Gordon L, Molina M, Ramanathan V, Rockström J, Scheffer M, Schellnhuber HJ, Svedin U. The anthropocene: from global change to planetary stewardship. *Ambio*. 2011 Nov;40(7):739-61. DOI: 10.1007/s13280-011-0185-x. PMID: 22338713; PMCID: PMC3357752.

Székely, M., Carletto, L. and Garami, A. (2015) The pathophysiology of heat exposure. *Temperature* 2, 452–452.

Tan, J., Zheng, Y., Tang, X., Guo, C., Li, L. *et al*. (2010) The urban heat island and its impact on heat waves and human health in Shanghai. *International Journal of Biometeorology* 54, 75–84.

Tempesta, D., Socci, V., De Gennaro, L. and Ferrara, M. (2018) Sleep and emotional processing. *Sleep Medicine Reviews* 40, 183–195.

Thom, E.C. (1959) The discomfort index. *Weatherwise* 12, 57–60.

Trenberth, K.E. (1999) Conceptual framework for changes of extremes of the hydrological cycle with climate change. *Climatic Change* 42, 327–339.

Trenberth, K.E. (2011) Changes in precipitation with climate change. *Climate Research* 47, 123–138.

Vallianou, N.G., Geladari, E.V., Kounatidis, D., Geladari, C.V., Stratigou, T. *et al*. (2021) Diabetes mellitus in the era of climate change. *Diabetes & Metabolism* 47, 101205.

Vardoulakis, S., Dimitroulopoulou, C., Thornes, J., Lai, K.M., Taylor, J. *et al*. (2015) Impact of climate change on the domestic indoor environment and associated health risks in the UK. *Environment International* 85, 299–313.

Vardoulakis, S., Matthews, V., Bailie, R.S., Hu, W., Salvador-Carulla, L., Barratt, A.L., Chu, C. Building resilience to Australian flood disasters in the face of climate change. *Med J Aust*. 2022 Oct 3;217(7):342-345.

Voogt, J.A. (2002) Urban heat island. In: Munn, T. (ed.) *Encyclopedia of Global Environmental Change*. Wiley, New York, pp. 660–666.

Walker, G. and Day, R. (2012) Fuel poverty as injustice: Integrating distribution, recognition and procedure in the struggle for affordable warmth. *Energy Policy* 49, 69–75.

Wang, J.X.L. and Gaffen, D.J. (2001) Late-twentieth-century climatology and trends of surface humidity and temperature in China. *Journal of Climate* 14, 2833–2845.

Wang, C., Liu, K. and Wang, H. (2023) The effects of temperature on sleep experience: Evidence from China. *Applied Economics* 55, 4678–4694.

Wei, D., Zhao, G., Liu, S. and Yang, L. (2022) Indoor thermal comfort in a rural dwelling in Southwest China. *Frontiers In Public Health* 10, 1029390.

Whitmee, S., Haines, A., Beyrer, C., Boltz, F., Capon, A.G. *et al*. (2015) Safeguarding human health in the Anthropocene epoch: Report of the Rockefeller Foundation–Lancet Commission on planetary health. *The Lancet* 386, 1973–2028.

Widerynski, S., Schramm, P., Conlon, K., Noe, R., Grossman, E. *et al*. (2017) *The Use of Cooling Centers to Prevent Heat-Related Illness: Summary of Evidence and Strategies for Implementation*. Climate and Health Technical Report Series, Climate and Health Program, Centers for Disease Control and Prevention National Center for Environmental Health, Atlanta, Georgia.

WMO (2024) Climate change indicators reached record levels in 2023: WMO. Press release. Available at: https://wmo.int/news/media-centre/climate-change-indicators-reached-record-levels-2023-wmo#:~:text=The%20WMO%20report%20confirmed%20that,ten%2Dyear%20period%20on%20record (accessed 11 July 2025).

WMO 2025. WMO Confirms 2024 as Warmest Year on Record at About 1.55°C Above Pre-Industrial Level. Geneva, Switzerland: The World Meteorological Organization.

Wouters, H., De Ridder, K., Poelmans, L., Willems, P., Brouwers, J. *et al*. (2017) Heat stress increase under climate change twice as large in cities as in rural areas: A study for a densely populated midlatitude maritime region. *Geophysical Research Letters* 44, 8997–9007.

Yao, R., Li, B. and Liu, J. (2009) A theoretical adaptive model of thermal comfort – Adaptive Predicted Mean Vote (APMV). *Building And Environment* 44, 2089–2096.

Yun, G.Y. (2018) Influences of perceived control on thermal comfort and energy use in buildings. *Energy & Buildings* 158, 822–830.

Zou, M. and Zhang, H. (2021) Cooling strategies for thermal comfort in cities: A review of key methods in landscape design. *Environmental Science and Pollution Research* 28, 62640–62650.

# 7 Healthy, Resilient Cities

**Evelyne de Leeuw[1]* and Geoff Green[2]**

[1]*Canada Excellence Research Chair 'One Urban Health', University of Montreal, Canada;* [2]*Centre for Regional, Economic and Social Research, Sheffield Hallam University, UK*

**Abstract**

Cities are human settings that connect with the natural environment in unique ways, mediated by the social contexts of experience. They are also impacted by history, policy and governance and importantly are human adaptations to a myriad of factors. A question of importance is how city resilience might be conceived and implemented. This chapter will connect the human experience with the built and natural environment in a multidisciplinary appraisal of sustainability and resilience in the face of the climate change and weather variability.

## Introduction

More people than ever live in cities in the early 21st century. This trend will continue, so planet Terra will become a wholly urban entity. The urban scientist Doxiadis (1962) has called this the 'ecumenopolis'. In his view, such a planetary city would manage to balance thriving ecologies, man-made environments, at vast scales. Cities are human creations, Doxiadis believes, and humanity has an innate capability to do good and improve. Cities are carefully planned, zoned, shaped and managed. Any problems and challenges in cities, therefore, may be resolved through better human consideration and action. This thinking aligns itself well with the declaration of the 'Anthropocene' – the geological era in which humanity has made an indelible impact on the planet's geosphere. This impact ranges from layers of man-made plastics or soot, through permanent scars inflicted on the surfaces of the Earth through mining and war, and changes in the biosphere through deforestation, acidification etc. In fact, it extends already beyond the 'naturally' created biosphere into space (De Leeuw, 2025).

In this chapter, we will look at this presumed interactions between humans, geosphere and biosphere, and how this would create opportunities for healthy and resilient global settlements in the face of what some have called multi-crises or permacrises.

## Spheres of Health and Resilience

It makes some sense to set off on this journey by describing the broader global-local ('glocal') context that shapes our perspectives. This is pertinent, as in recent years there have been urgent calls to develop and cherish a more holistic approach to health that has been labelled 'planetary health'. There even are *Professors of Planetary Health* these days, a notion that must be cognitively challenging – imagine compressing

*Corresponding author: evelyne.de.leeuw@umontreal.ca

 *Climate Change and Risk Mitigation: Reducing Vulnerabilities and Enhancing Resilience* (ed. P. Barnes)
DOI: 10.1079/9781800624153.0007

all of the planet into one tiny brain or research team .... The Planetary Health Alliance defines its moniker as 'a solutions-oriented, transdisciplinary field and social movement focused on analyzing and addressing the impacts of human disruptions to Earth's natural systems on human health and all life on Earth' (Planetary Health Alliance, n.d.). Particularly considering climate change, this is an important distinction. The United Nations Framework Convention on Climate Change (UNFCC) similarly recognises that 'human health and the health of our planet are inextricably linked, and that our civilization depends on human health, flourishing natural systems, and the wise stewardship of natural resources' (UNFCC, 2023). At COP 28, concrete action on this statement was taken by, for the first time, recognising health as a critical element in the global deliberations.

For us, an ultimate question is whether human health should be at the core of these conceptualisations (e.g. Fig. 7.1).

There is a danger in centring everything on 'health'. In the socially critical 1970s, and as a corollary of feminist and environmental scholarship and social movements,

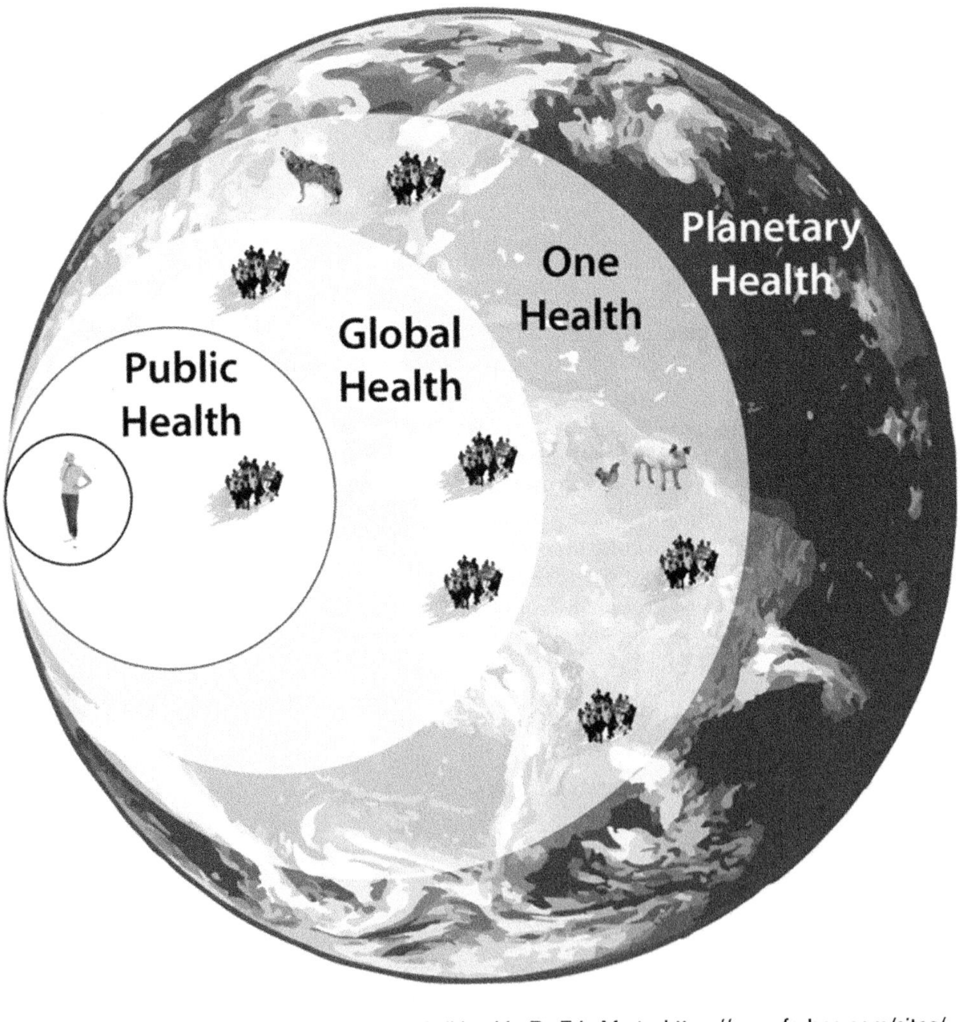


**Fig. 7.1.** Planetary Health as the outer shell of all health. By Eric Marty, https://www.forbes.com/sites/johndrake/2021/04/22/what-is-planetary-health/. This image has a Creative Commons license that allows for sharing.

Ehrenreich and Ehrenreich (1970) identified the existence of a 'medical–industrial complex' (inspired by Dwight D. Eisenhower's warning of a 'military-industrial complex' in his Presidential farewell speech). The sole purpose of this conglomerate of healthcare professionals (and particularly medical doctors), pharmaceutical and other medical technology industries, health service delivery institutions and disease care consumers (see De Leeuw, 2021) remains to maximise influence, power and profit, often at the expense of optimal and equitable health and wellbeing of individuals and communities, let alone the planet. One of the means of pursuing this has been deemed 'medicalisation' (e.g. Busfield, 2017) – bringing social phenomena under a medical regime, for example, normal life course events (such as childbirth or menopause) deemed in need of medical intervention. The flip side of medicalisation may be 'healthism' (Crawford, 1980), which extends the magic of medicalisation to other arenas, for instance, the food industry and agribusiness, education and transport. A particular form of 'healthism' has been practised since the 1980s by those that subscribe to the idea of 'Healthy Public Policy' or 'Health in All Policy', for example, De Leeuw (2013). With health being all-important, every sphere of life can and should be mobilised to maximise salutogenesis (Antonovsky and Sagy, 2022 – 'the making of health', as opposed to pathogenesis, 'disease creating'). In a critical analysis of the potential of inserting health in all (policies or planet), De Leeuw (2017) finds that appealing to health considerations has counterproductive effects for the prominence of the 'health' discourse (frequently merely the disease discourse). A survey of European health policy makers identified them claiming that we need to 'Avoid the H word' (with H indicating 'Health'): these policy operators say that the development of whole-of-government policies for health and wellbeing should not start with health claims, as potential contributors to such integral notions would switch off from the policy process. Their responses would include them identifying the dominance and prominence of the health sector (i.e. the above 'medical–industrial complex') and showing great hesitation in investing in a further strengthening of its power through joined-up policies for health.

To build healthy and resilient cities in the face of global (climate, but also other) change it quite likely makes sense to avoid centring the argument on health, but rather to focus on life on our planet more broadly. In Fig. 7.2 we suggest several concentric spheres. These start with the 'hardware' of our – and any – planet, the geosphere. The geosphere is enveloped in our watery (hydrosphere) and airy (atmosphere) confines. It took about a billion years for these spheres to generate the first traces of life, and another billion before stromatolites and cyanobacteria started filling the atmosphere with compounds that would sustain more complex life forms. This all shaped a 'biosphere' – the layers of living and complex ecosystems. One of the newest life forms on Terra, Homo (currently *sapiens*) subsequently, and extremely rapidly by the clock of planetary evolution, put its stamp on these 'outer' spheres. In its 'anthroposphere', humanity pretended to have dominion over all the world – an illusion sometimes presumed given by supernatural beings. Through its dominion over geosphere and biosphere we, *Homo sapiens*, have become pedantic and filled with hubris. But the unique proclivities of the anthropogenic cosmology (science, creativity, art, belief systems) may (and should) also well provide impetus for fixing past trespasses.

## Resilience and Human Action

This chapter isn't necessarily the crux in unequivocally defining resilience, and in particular 'urban health resilience'. But it makes sense to momentarily dwell on scope. Definitions of (urban) resilience are plentiful (Meerow *et al.*, 2016) and we take our bearings for health resilience from Iwaniec *et al.* (2021) and a landmark report from the Public Health Agency of Canada (2023). The latter states, 'A resilient community, whether defined by geography, interest, experience, or identity, is one that has the resources to help prevent, withstand, and recover from emergencies. Resilient communities are able to adapt despite disturbances caused by

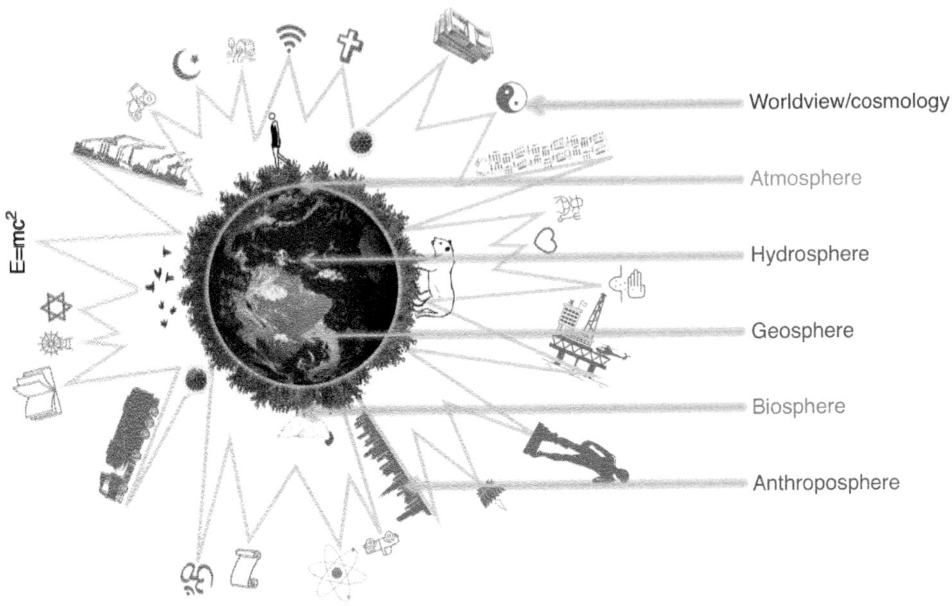


**Fig. 7.2.** The spheres of life, our planet and resilience. Authors' own image.

emergencies and return to acceptable levels of functioning. Further, the community attributes and systems that support emergency resilience could also contribute to broader social, economic, health, and environmental benefits before, during, and after emergencies' (2023, p. 8). The former grounds itself in the disciplinary foundations of ecology, and sees resilience as the 'capacity of a system to maintain its basic structure, function, and identity while undergoing change in the face of shocks and stresses' (2021, p. 5). Bringing these and similar thoughts together, Miller *et al.* (2024) simply – and elegantly – frame resilience as a transdisciplinary form of responsive thinking that recognises environmental, economic, ecological and social dynamics.

A critical difference between a health viewpoint and Miller *et al.*'s ecological viewpoint for resilience is that the health community recognises the differential and often inequitable impacts of adverse health events and emergencies. Whereas an ecological perspective may simply indicate a 'bouncing back' without much further analysis into causes and mechanisms of impact, a health gaze recognises the importance of resilience as integral to social community infrastructure (Poland *et al.*, 2021). Relating this particular flavour of infrastructure to the very nature of urban (social; environmental; morphological and design; cultural) environments creates a much more urgent and responsive perspective for urban *health* resilience. Add to this the looming disaster of climate change, and its wide-ranging differential, distributed, inequitable and permanent impacts, and it is clear that a resilient response to urban climate change is urgent.

Health resilience is a particular domain within the broad and exuberant resilience literature. Some authors, including Döring *et al.* (2015), see that health and resilience are inseparable and reciprocal. Without (human and urban) health, no resilience, and the other way around. An important perspective that they codify is that we need to distinguish more operationally between the three ingredients of the resilience mechanism: disturbance – response – outcome. Each of these can be imagined and scaled, for instance, for intensity, priority or severity (see Fig. 7.3).

In cities, which are constructed human-urban imaginings and realities (see Barton, 2005, Fig. 7.4), one can envision that the ever-increasing complex nature of the places we live in, how they impact on all – proximal

**Fig. 7.3.** Mapping of resilience scales and dimensions. Derived from Döring *et al.*, 2015.

and distal – determinants of health, as well as the epistemic, spatial and social injustices that are the result of perverse differences between populations and livelihoods, creates imagined limitations to the responses on the vertical axis in Fig. 7.3 above.

Fortunately, city dwellers, their communities and institutional arrangements (including local political leadership and entrepreneurship) have responded to these increasing glocal complexities. Whereas 'local government' used to be merely considered a lower level of public sector governance, it is now increasingly framed as a critical interface of the local and the global – be it for humans, ecosystems and biodiversity or cultural expression (e.g. Pejic *et al.*, 2024).

## Cities Get Organized

The World Health Organization (WHO) Healthy Cities Network was conceptualised in the mid-1980s during a think-tank in Lisbon (in 1985) with social movement and urban health experts and activists, and a year later in Dusseldorf with a small initial group of European cities. The European office of WHO then invited cities to nominate themselves to become a member of this experimental endeavour that was to be one of the delivery vehicles of the Ottawa Charter for Health Promotion. The enthusiasm of European cities for the initiative was overwhelming, and WHO was stimulated to undertake forms of vetting and designation of 'official' Healthy Cities and Healthy City Networks (De Leeuw, 2022a). The latter coagulated around national interests, language (French and German) and particular interests (research being a prominent one). Beyond the European Region of WHO, many other initiatives quickly followed suit, in Canada, Australia, Japan and New Zealand. The enthusiasm and commitment of local governments for health has never ceased – in fact it diversified, to now include a range of diverse global networks that embrace urban health (Kim *et al.*, 2022).

Arguably, the Healthy Cities Network established in the 1980s was – and continues to be – the first multilateral and transnational network of local governments, not just for health, but in general. They preceded the networks of local governments that were prompted by the 1992 Rio de Janeiro Sustainability Conference (through the International Council for Local Environmental Initiatives, ICLEI) and transcended the much less committal twin or sister cities approach that emerged upon World War II.

In the pack that followed Healthy Cities we have seen the emergence of a multitude of

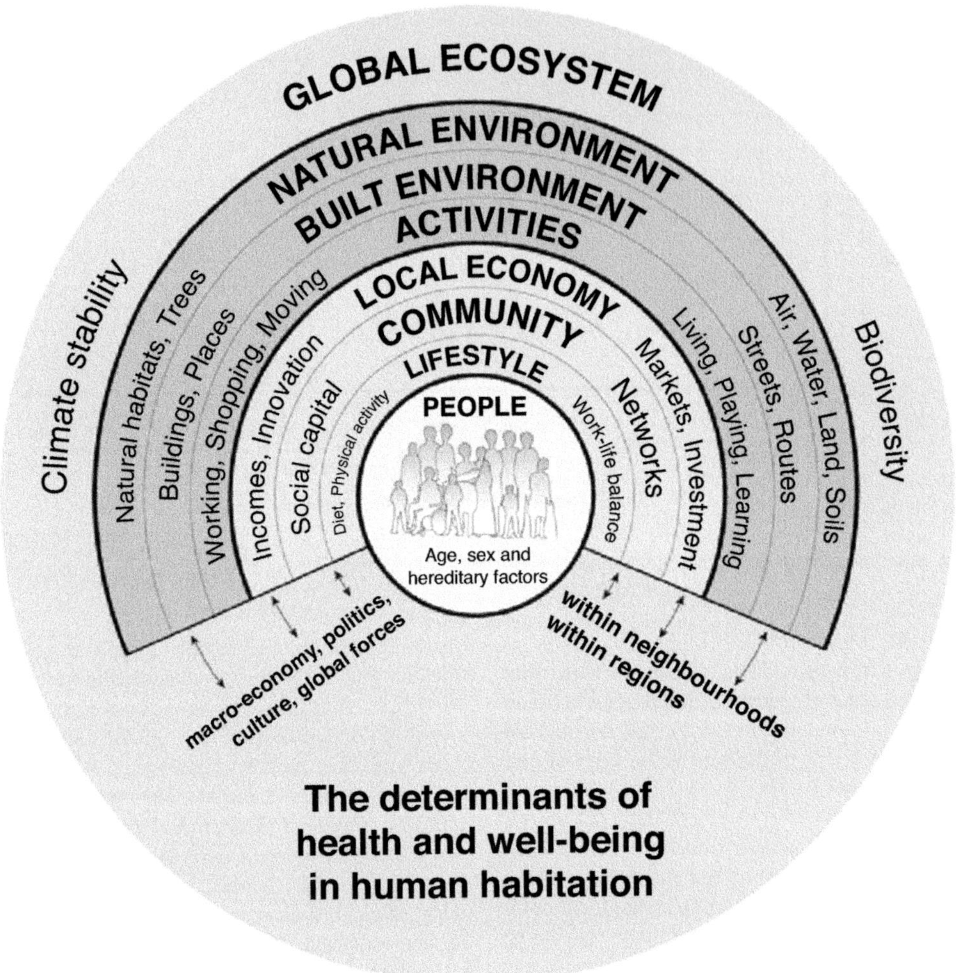


**Fig. 7.4.** Settlement health map. From Barton and Grant, 2006, developed from a concept by Dahlgren and Whitehead, 1991. Used with permission.

'Theme Cities' (Forbat *et al.*, 2023). They include the perhaps well-known Age-Friendly Cities, but also networks around Resilient Cities, Smart Cities and Sustainable Cities. Other Theme Cities are the perhaps more intriguing (and possibly more prominent) Slow Cities, Winter Cities, Festival Cities and Transition Towns. Health, wellbeing and resilience are all notions integral to these. At the local level, momentum is building to create, strengthen and sustain better futures in the face of 'polycrises' and 'permacrises' (De Leeuw, 2022b).

## Resilience in the Face of Crisis

Cities are also recognised in global discourse as key to planetary survival and human thriving. In Table 7.1 we summarise developments in recent decades where the local and urban levels are explicitly recognised, including in the context of the Sustainable Development Goals and Climate Change.

Of particular note in this list is the presence of the UNFCCC – the parent treaty and

**Table 7.1.** Statements and developments from global gatherings and commitments to resilience and planetary survival, including climate change and sustainability. Authors' own table.

| Year | Event | Resilience? | Cities/local government focus |
|---|---|---|---|
| **1992** | Agenda 21, Rio Declaration | ***9.19*** *'The loss of biological diversity may reduce the* ***resilience*** *of ecosystems to climatic conditions.'* ***12.19*** *new practices to 'reflect the* ***resilience*** *of the whole ecological and social system.'* ***35.2*** *'This is essential if a more accurate assessment is to be provided for the carrying capacity of planet Earth and its* ***resilience*** *under the many stressors placed upon it by human activities.'* | **Chapter 7: PROMOTING HUMAN SETTLEMENT DEVELOPMENT** ***7.1*** *'In industrialized countries, the consumption patterns of cities are severely stressing the global ecosystem ...'* ***7.20*** *'Some international initiatives in support of such efforts, as in the Sustainable Cities Programme of Habitat and the Healthy Cities Programme of WHO, should be intensified.'* **Chapter 28: LOCAL AUTHORITIES' INITIATIVES IN SUPPORT OF AGENDA 21.** ***'28.1*** *Because so many of the problems and solutions being addressed by Agenda 21 have their roots in local activities, the participation and cooperation of* ***local authorities*** *will be a determining factor in fulfilling its objectives. Local authorities construct, operate and maintain economic, social and environmental infrastructure, oversee planning processes, establish local environmental policies and regulations, and assist in implementing national and subnational environmental policies. As the level of governance closest to the people, they play a vital role in educating, mobilizing and responding to the public to promote sustainable development.'* |

*Continued*

**Table 7.1.** Continued

| Year | Event | Resilience? | Cities/local government focus |
|---|---|---|---|
| **2015** | United Nations Framework Convention on Climate Change (UNFCCC)<br>COP 21 (21st meeting of the Conference of Parties)<br>*Paris Agreement* | ***Article 7***<br>*'Parties* (National Governments) *hereby establish the global goal on adaptation of enhancing adaptive capacity, strengthening* ***resilience*** *and reducing vulnerability to climate change, with a view to contributing to sustainable development and ensuring an adequate adaptation response in the context of the temperature goal referred to in Article 2.'* | ***Article 11***<br>*Foster 'ownership at the national, subnational and local levels.'* |
| **2015** | United Nations<br>New York Summit on Sustainable Development Goals (SDG)<br>*Adopted*<br>2030 Agenda for Sustainable Development | *SDG Goal 11.*<br>*Make cities and human settlements inclusive, safe,* ***resilient*** *and sustainable.*<br>11b. *By 2020, substantially increase the number of cities and human settlements adopting and implementing integrated policies and plans towards inclusion, resource efficiency, mitigation and adaptation to climate change,* ***resilience*** *to disasters …*<br>*13.1 Strengthen* ***resilience*** *and adaptive capacity to climate-related hazards and natural disasters in all countries.*<br>*Ref in preamble to 'People. Planet, Prosperity, Peace and Partnership' later used as motif for Phase VII of Healthy Cities Network* | *34. We recognize that sustainable urban development and management are crucial to the quality of life of our people. We will work with local authorities and communities to renew and plan our cities and human settlements so as to foster community cohesion and personal security and to stimulate innovation and employment.*<br>This, above, is the only substantial reference to local government but then big push for recognition by United Cities and Local Gvt. As related to all 17 Goals. Led to UN platform *Local2030: Localising the SDGs. https://www.local2030.org/library/view/705* |

*Continued*

**Table 7.1.** Continued

| Year | Event | Resilience? | Cities/local government focus |
|---|---|---|---|
| **2016** | United Nations Framework Convention on Climate Change (UNFCCC)<br>COP 22<br>Marrakech<br>CMA =<br>Technical body serving parties to the Paris Agreement | Global Climate Action<br>Sustainable Development Goals adopted in 2015 | Multi-level Governance of UN Global response to Climate Change.<br>*'Since adoption of the Paris Agreement, global climate action has been encouraged and facilitated under the banner of the Marrakech Partnership for Global Climate Action, which was agreed in Morocco at COP 22 and acknowledged at subsequent Conferences of the Parties (CMA). The (Marrakech) Partnership brings together stakeholders working in key sectors and themes to spur enhanced climate ambition and action, and then recognizes that action, to inspire still greater effort.'* |
| | | *'Marrakech, November 14, 2016 – Today, during the 22nd Conference of the Parties (COP22) to the UN Framework Convention on Climate Change (UNFCCC) in Marrakech,* ***local leaders*** *from 7100 cities announced the collective impact of their actions to accelerate world progress on meeting the commitments of the Paris Agreement and stressed the importance of expanding access to city-climate finance. By taking bold action and tracking their progress transparently, the cities of the soon-to-be Global Covenant of Mayors for Climate & Energy — a coalition created by the Compact of Mayors and EU Covenant of Mayors — are demonstrating what can be accomplished when cities are empowered and supported to reduce emissions, build* ***resilience*** *and work collaboratively'* (Global Covenant of Mayors for Climate & Energy, 2016). | |
| **2023** | COP 28<br>UAE | Main COP 28 UAE DECARATION ON CLIMATE, RELIEF, RECOVERY AND PEACE<br>Commitment to *'Enhanced financial support for climate adaptation and* ***resilience****.'*<br>Via technical support from CMAs (1–5) and COPs 23–27 (Sindy, 2023) and learning the lessons of COVID-19, a key agreement is the adoption of the UAE FRAMEWORK FOR GLOBAL CLIMATE **RESILIENCE**. | |

*Continued*

**Table 7.1.** Continued

| Year | Event | Resilience? | Cities/local government focus |
|---|---|---|---|
| | | *The GLOBAL GOAL OF ADAPTATION* 9e. *'Increasing the resilience of infrastructure and human settlements to climate change impacts to ensure basic and continuous essential services for all, and minimizing climate-related impacts on infrastructure and human settlements'.* | Main Declaration '4. *Prioritizing local ownership, impact, and results where possible, including through channelling finance at the local level to respond to local needs and priorities and working with affected communities and both local government and non-government partners'.* |

driver of the world's commitments to address the climate change emergency. All of the above global developments have *local* drivers and consequences, and an ever-increasing evidence base. But the 2023 UNFCCC Conference of Parties was not only the first to host a Health Day and Health Pavilion, but was also the one where the Intergovernmental Panel on Climate Change (IPCC) devoted an entire volume to the impacts of climate change on human, ecological and planetary health (Lee *et al.*, 2023). The arguments and graphics are, even after a brutal process of scholarly and political peer review, terrifying. For instance, the scenarios for the spread of *Aedes aegypti* (the mosquito associated with the spread of dengue fever, chikungunya, Zika fever, Mayaro and yellow fever viruses and thus for a very significant burden of disease) are shocking (Fig. 7.5).

Chapter 7 of the IPCC report integrates general health, wellbeing and community considerations with the existing threats (and possible, resilient opportunities) as well as scenarios related to living in urban environments and an ever-further urbanising world. The *Aedes aegypti* figure (Fig. 7.5) illustrates the urban doom scenario. Not only is the mosquito escaping the jungles that were traditionally its habitat to invade cities in the tropics, but also it starts occupying urban agglomerations that never had to be prepared for an invasion and attack of mosquito-borne diseases. This is perhaps most vividly illustrated by zooming in on the south-east periphery of Australia – where traditionally 90% of the Australian population lives. These are urban populations within urban socio-ecological and political environments that never had to endure such a mosquito onslaught (Webb, 2020), and where one can doubt their resilience capacity.

Intriguingly, the IPCC Health report, when focusing on solutions, fails to embrace an integrated urban agenda (admittedly, a special IPCC Cities report is currently being commissioned, cf. Bai, 2023). Instead, the Health report regresses to a more traditional healthcare system (i.e. disease driven) perspective, following the lead of the World Health Organization's climate resilient health systems framework (Fig. 7.6; WHO, 2015a).

Surprisingly, the urban dimension to healthcare climate resilience is no longer front and centre to WHO's development agenda. It seems as if healthcare systems are autonomous entities (perhaps to be called 'precincts' – see De Leeuw, 2023) separate from any urban integration and embeddedness. Such a perspective, real or imagined, does not bode well for urban health resilience parameters. It fails to acknowledge the ecological primacy, and may not take into account extremely relevant innovation and social development around nature-based solutions for cities (including wellbeing in all Fig. 7.2 spheres, cf. Frantzeskaki *et al.*, 2019), architecture and design (Thomas and Da Cunha, 2017; Daniel-Lacombe, 2023) and creative transdisciplinarity (that is, working with systems and communities in a joint solutions-driven process of discovery (Steele *et al.*, 2023).

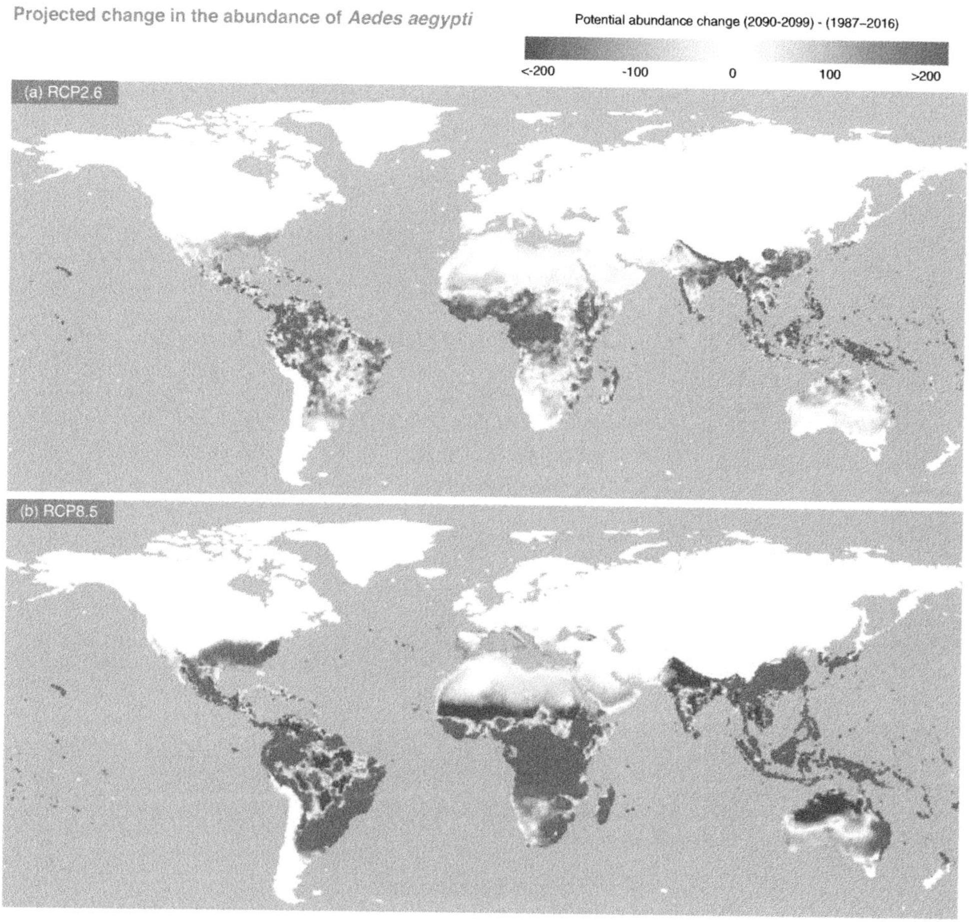


**Fig. 7.5.** Projected spread (into the world's urban areas) of a type of mosquito (*Aedes aegypti*) that is a vector for 54 viruses and diseases such as dengue fever, chikungunya, Zika fever, Mayaro and yellow fever. From Sarkar *et al.*, 2019. Published under an open access Creative Common CC BY license.

## Urban Governance for Resilience

Understanding the challenges to urban environments posed by the polycrisis that include the climate change emergency requires an analysis of the local governance parameters in each unique situation. This is a key premise of realist analysis, where 'context' is explicitly taken into account when reviewing policy processes and their outcomes (Pawson and Tilley, 2004). This means that varying contexts are no longer uncontrollable black boxes that stand in the way of standardised comparisons, but that they can be described and framed for each situation. If we see 'governance' as a supreme driver of context, it is important to describe it diligently and specifically. Theory of governance (which can simply be defined as 'how we do things around here' and includes the specification of levels of implicit and explicit rules) would then allow for the description of different patterns of governance for each situation, time and environment (e.g. Hill and Hupe, 2006; Hupe and Hill, 2006). Green was the first to attempt to provide a systematic (and systemic) overview of health governance for each of these, with a focus of various levels of government of member states of the European Region of WHO (Green, 1998). The work – in need of critical updates in a newly challenged world – shows how

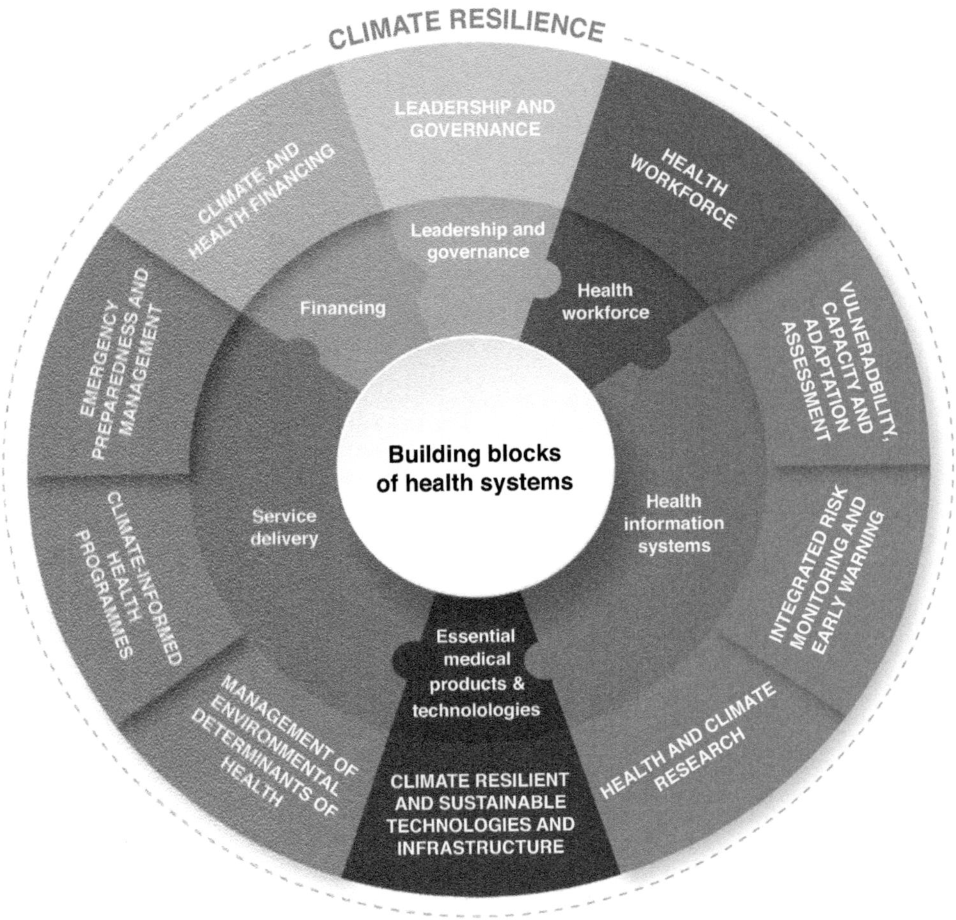


**Fig. 7.6.** Ten components of the WHO operational framework for building climate resilient health systems, with links to the building blocks of health systems. From WHO, 2015a.

responsibilities, institutional and professional agency, and rules for the delivery of health and prevention are covering an enormous array of variability. In some countries, local governments are wholly responsible for the planning and delivery of health, prevention and resilience mechanisms (but often not for generating the income, and distribution of finance...!) whereas in others there is a complex hierarchy in which central decision making attempts to analyse and then devolve the required resources, rules and interventions (e.g. Hoeijmakers, 2005, describes for The Netherlands how local epidemiologists are expected to shape policy for health promotion – a legislative attempt that was corrected by recognising that policy making is not necessarily an epidemiologist's prerogative (see Cairney and St Denny, 2020).

The practical consequence of this distributed reality can be seen in the guidance that has been developed over many years in the networks of age-friendly cities around the world, and specifically in the WHO and European Network AFEE (Age-Friendly Environments in Europe; WHO and EU, 2017). Figure 7.7 shows the key governance domains for creating, sustaining, evaluating and adapting age-friendly environments. These eight components of age-friendly urban environments framed cities' responses to the COVID-19 pandemic (WHO, 2024) and would provide evidence-based directions for climate change-resilient urban development,

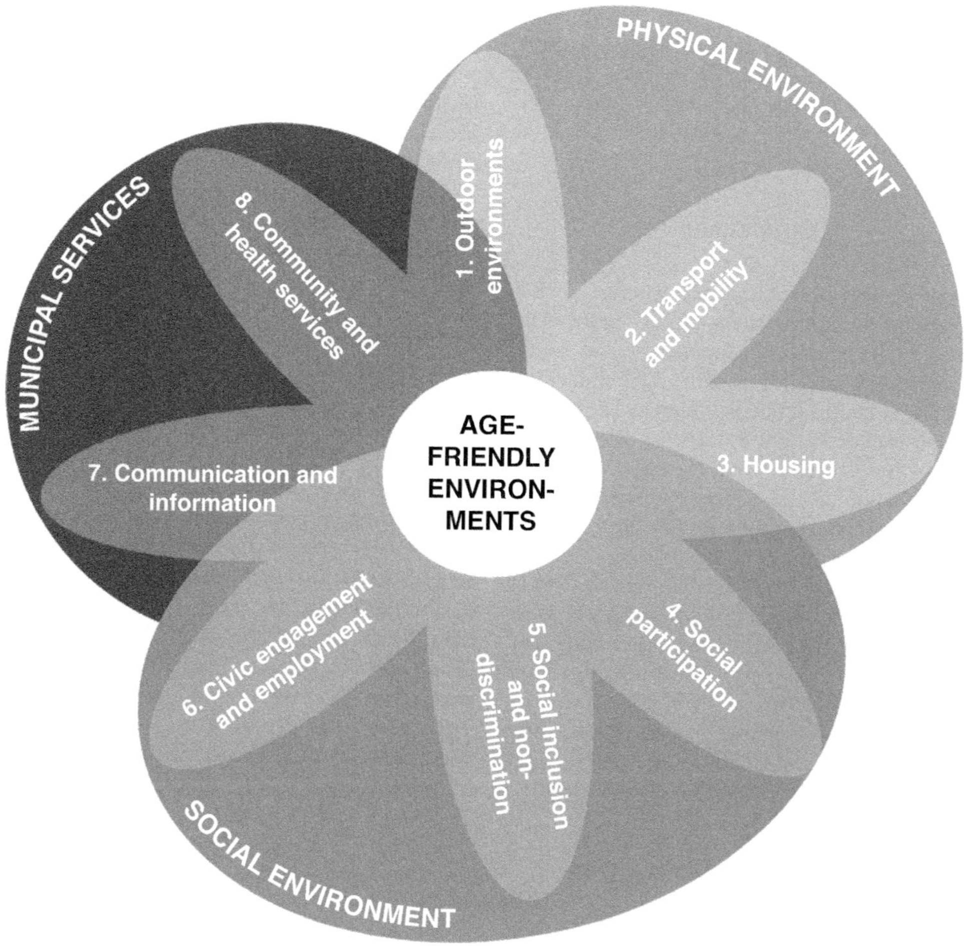


**Fig. 7.7.** Framework of the AFEE project. From WHO and EU, 2017.

too. In the AFEE context and drawing on the Resilient Cities Network, urban resilience is defined as 'the capacity of a city's systems, businesses, institutions, communities and individuals to survive, adapt and thrive, no matter what chronic stresses and acute shocks they experience' (Resilient Cities Network, 2023). Holman and Walker (2021) record how over the life course, older people can use their individual agency to capitalise on these resilient, age-friendly environments.

The WHO (2015b) *World Health Report on Ageing and Health* embraces a public health framework for healthy ageing, which conceptualises an individual's resilience as the ability to maintain or improve a level of functional ability in the face of adversity (through resistance, recovery or adaptation). City responses to COVID-19 revealed how investment in the physical environment – in the public realm of parks and neighbourhood streets – encouraged individual and community distancing, a form of 'survival resilience'. Investment in cities' social infrastructure enabled 'adaptive resilience' – community cohesion, trust and social capital strengthening individuals' mental capabilities in the post-pandemic 'new normal'. This may appear to be, at first sight, a behaviourist approach of complex systemic challenges, but within the multi-level governance framework that we outlined above, it will be clear that creating urban resilience is a joined-up endeavour that can start at many levels, from individual agency to institutional responsibility and constitutional parameters. An urban focus on climate change resilience, we feel, has a lot to learn from an integrated age-friendly place-based – and proven – approach.

## Conclusion

Cities are settings for and recipients of cascading global forces; they also have a 15,000-year history of constant adaptation (and more recently, uncontrolled growth). The challenges to urban (health) resilience emanate either from nested ecological spheres latterly disrupted by the Anthropocene, or from the political ecosystems of national and international governance summarised in UN frameworks. They are not merely the expression of demographic change and urbanisation, cities now being the habitat for over half the world's population. Yes, context is critical, eroding or enhancing urban and urbanising environments' capacity to secure a resilient future. Yet, at their existential core is agency and democratic accountability – city leadership, by politicians and intersectoral partners, local and at every level of government. A Task Force of 16 European cities summarised well a first lesson from their collective appraisal of the COVID-19 pandemic – *Build Resilience Together*: City leadership during the coronavirus disease (COVID-19) pandemic was nested within innovative systems and structures combining many sectors and all levels of government. It is vital to utilize this ecosystem to prepare for and respond to future health emergencies (WHO, 2024).

If there is *one* lesson that emerges from our account of thinking about urban health resilience, it is that we need to break free from Cartesian disciplinary atomisation, hyperspecialisation towards technocratic approaches of planetary and ecological issues, and 'expert'-driven solutions. Even when the latter allude to all-important approaches such as citizen science or knowledge co-creation (often subverted to denote merely 'cheap' ways of data collection...), it is important to recognise that humans are not the centre of all (De Leeuw, 2022c, 2025). Urban resilience for health and wellbeing in the face of the polycrisis can and should be much more integrated and engaged (WHO, 2022) – and this is not a pipedream, as the work of the AFEE has shown.

## References

Antonovsky, A. and Sagy, S. (2022) Aaron Antonovsky, the scholar and the man behind salutogenesis. In: Mittelmark, M.B., Bauer, G.F., Vaandrager, L., Pelikan, J.M., Sagy, S. *et al.* (eds) *The Handbook of Salutogenesis*. Springer, Cham, Switzerland, pp. 15–23.

Bai, X. (2023) Make the upcoming IPCC Cities Special Report count. *Science* 382(6670), eadl1522.

Barton, H. (2005) A health map for urban planners. *Built Environment* 31(4), 339–355.

Barton, H. and Grant, M. (2006) A health map for the local human habitat. *The Journal for the Royal Society for the Promotion of Health* 126(6), 252–253.

Busfield, J. (2017) The concept of medicalisation reassessed. *Sociology of Health & Illness* 39(5), 759–774.

Cairney, P. and St Denny, E. (2020) *Why Isn't Government Policy More Preventive?* Oxford University Press, Oxford, UK.

Crawford, R. (1980) Healthism and the medicalization of everyday life. *International Journal of Health Services* 10(3), 365–388.

Dahlgren, G. and Whitehead, M. (1991) The main determinants of health model. In: Dahlgren, G. and Whitehead, M. (eds) *A Discussion Paper on European Strategies for Tackling Social Inequities in Health*. WHO Regional Office for Europe, Copenhagen. Available at: https://iris.who.int/handle/10665/107791

Daniel-Lacombe, É. (2023) *Vers une architecture pour la santé du vivant*. Les Presses de l'Université de Montréal, Montreal, Canada.

De Leeuw, E. (2013) The public policy approach: Governments, institutions, welfare states and social justice. In: McQueen, D.V. (ed.) *Global Handbook on Noncommunicable Diseases and Health Promotion*. Springer, New York, pp. 237–248.

De Leeuw, E. (2017) Engagement of sectors other than health in integrated health governance, policy, and action. *Annual Review of Public Health* 38, 329–349.

De Leeuw, E. (2021) The rise of the consucrat. *International Journal of Health Policy and Management* 10(4), 176.

De Leeuw, E. (2022a) Researching a diverse epistemic social movement: The challenges and rewards of European healthy cities realist synthesis. In: Potvin, L. and Jourdan, D. (eds) *Global Handbook of Health Promotion Research, Vol. 1: Mapping Health Promotion Research*. Springer International Publishing, Cham, Switzerland, pp. 739–757.

De Leeuw, E. (2022b) Is local better? Evolving hybrid theorising for local health policies. In: Fafard, P., Cassola, A. and De Leeuw, E. (eds) *Integrating Science and Politics for Public Health*. Palgrave Studies in Public Health Policy Research. Palgrave Macmillan, Cham, Switzerland, pp. 153–185.

De Leeuw, E. (2022c) One Urban Health. URBANET - News and Debates on Municipal and Local Governance, Sustainable Urban Development and Decentralisation. GIZ, German Federal Ministry for Economic Cooperation and Development (BMZ), Bonn. Available at: https://www.urbanet.info/one-urban-health/ (accessed 16 July 2025).

De Leeuw, E. (2023) Precincts and promises. In: Biloria, N., Sebag, G. and Robertson, H. (eds) *The Empathic City. An Urban Health and Wellbeing Perspective*. Springer, Cham, Switzerland, pp. 347–365.

De Leeuw, E. (2025) Health beyond borders: The future of health promotion. *Scandinavian Journal of Public Health* 5(1). DOI: 10.1177/14034948241288272.

Döring, T.F., Vieweger, A., Pautasso, M., Vaarst, M., Finckh, M.R. *et al*. (2015) Resilience as a universal criterion of health. *Journal of the Science of Food and Agriculture* 95(3), 455–465.

Doxiadis, C.A. (1962) Ecumenopolis: Toward a universal city. *Ekistics* 13(75), 3–18.

Ehrenreich, B. and Ehrenreich, J. (1970) *The American Health Empire: Power, Profits and Politics*. Vintage Books, New York.

Forbat, J., De Leeuw, E. and Simos, J. (2023) Theme Cities networks: A typology. In: Brears, R.C. (ed.) *The Palgrave Encyclopedia of Urban and Regional Futures*. Springer International Publishing, Cham, Switzerland, pp. 1827–1832.

Frantzeskaki, N., McPhearson, T., Collier, M.J., Kendal, D., Bulkeley, H. *et al*. (2019) Nature-based solutions for urban climate change adaptation: Linking science, policy, and practice communities for evidence-based decision-making. *BioScience* 69(6), 455–466.

Global Covenant of Mayors for Climate & Energy (2016) Global Covenant of Mayors for Climate & Energy Announces Its Global Impact. Available at: https://www.globalcovenantofmayors.org/press/at-cop22-local-leaders-showcase-their-commitment-to-accelerating-global-progress-on-climate-change-2/ (accessed 16 July 2025).

Green, G. (1998) Health and Governance in European Cities: A Compendium of Trends and Responsibilities for Public Health in 46 Member States of the WHO European Region. European Hospital Management Journal Ltd, London.

Hill, M. and Hupe, P. (2006) Analysing policy processes as multiple governance: Accountability in social policy. *Policy & Politics* 34(3), 557–573.

Hoeijmakers, M.J.A. (2005) Local health policy development processes: Health promotion and network perspectives on local health policy-making in the Netherlands. PhD thesis, Maastricht University, Netherlands.

Holman, D. and Walker, A. (2021) Understanding unequal ageing: Towards a synthesis of intersectionality and life course analyses. *European Journal of Ageing* 18, 239–255. DOI: 10.1007/s10433-020-00582-7.

Hupe, P.L. and Hill, M.J. (2006) The three action levels of governance: Re-framing the policy process beyond the stages model. In: Peters, B.G. and Pierre, J. (eds) *Handbook of Public Policy*. Sage Publications, London, pp. 13–30.

Iwaniec, D.M., Grimm, N.B., McPhearson, T., Berbés-Blázquez, M., Cook, E.M. *et al*. (2021) A framework for resilient urban futures. In: Hamstead, Z.A., Iwaniec, D.M., McPhearson, T., Berbés-Blázquez, M., Cook, E.M. et al. (eds) *Resilient Urban Futures*. Springer Nature, Cham, Switzerland, pp. 1–9.

Kim, J., De Leeuw, E., Harris-Roxas, B. and Sainsbury, P. (2022) Four urban health paradigms: The search for coherence. *Cities* 128, 103806.

Lee, H., Calvin, K., Dasgupta, D., Krinner, G., Mukherji, A. *et al*. (2023) *Climate change 2023: Synthesis report, summary for policymakers*. Contribution of Working Groups I, II and III to the Sixth Assessment Report of the Intergovernmental Panel on Climate Change. IPCC, Geneva, Switzerland.

Meerow, S., Newell, J.P. and Stults, M. (2016) Defining urban resilience: A review. *Landscape and Urban Planning* 147, 38–49.

Miller, D., AlArasi, H. and Chapple, K. (2024) Special issue on climate justice and resilience. *Journal of City Climate Policy and Economy* 2(3), 269–270.

Pawson, R. and Tilley, N. (2004) *A Brief Introduction to Realist Evaluation*. UK Cabinet Office, London. Available at: https://assets.publishing.service.gov.uk/media/60f7fdf7d3bf7f56824cc634/Brief_introduction_to_realist_evaluation.pdf (accessed 16 July 2025).

Pejic, D., Leavesley, A., Matsumoto, T. and Acuto, M. (2024) National urban politics and policy: Between local, national, and global. In: Vogel, R.K. (ed.) *Handbook of Urban Politics and Policy*. Edward Elgar Publishing, Cheltenham, UK, pp. 405–420.

Planetary Health Alliance (n.d.) What is Planetary Health? Available at: https://planetaryhealthalliance.org/what-is-planetary-health/ (accessed 18 July 2025).

Poland, B., Gloger, A., Morgan, G.T., Lach, N., Jackson, S.F. *et al*. (2021) A connected community approach: Citizens and formal institutions working together to build community-centred resilience. *International Journal of Environmental Research and Public Health* 18(19), 10175.

Public Health Agency of Canada (2023) *Chief Public Health Officer of Canada's Report on the State of Public Health in Canada 2023: Creating the Conditions for Resilient Communities: A Public Health Approach to Emergencies*. Public Health Agency of Canada, Ottawa, Ontario.

Resilient Cities Network (2023) What Is Urban Resilience? Resilient Cities Network. Available at: https://resilientcitiesnetwork.org/what-is-urban- resilience/ (accessed 5 November 2025).

Sarkar, S., Gangare, V., Singh, P. and Dhiman, R.C. (2019) Shift in potential malaria transmission areas in India, using the fuzzy-based climate suitability malaria transmission (FCSMT) model under changing climatic conditions. *International Journal of Environmental Research and Public Health* 16(18), 3474. DOI: 10.3390/ijerph16183474.

Sindy, S. (2023) The UAE Framework for Global Climate Resilience. Available at: https://www.linkedin.com/pulse/uae-framework-global-climate-resilience-sindy-singh-s71sf/ (accessed 16 July 2025).

Steele, W., Handmer, J. and McShane, I. (2023) *Hot Cities: A Transdisciplinary Agenda*. Edward Elgar Publishing, Cheltenham, UK.

Thomas, I. and Da Cunha, A. (2017) *La ville résiliente: Comment la construire?* Les Presses de l'Université de Montréal, Montreal, Canada.

UNFCC (2023) Planetary Health. Available at: https://unfccc.int/climate-action/un-global-climate-action-awards/planetary-health (accessed 18 July 2025).

Webb, C. (2020) Reflections on a highly unusual summer: Bushfires, COVID-19 and mosquito-borne disease in NSW, Australia. *Public Health Research & Practice* 30(4).

WHO (2015a) *Operational Framework for Building Climate Resilient Health Systems*. World Health Organization, Geneva, Switzerland. Available at: https://iris.who.int/bitstream/handle/10665/189951/9789241565073_eng.pdf?sequence=1 (accessed 14 July 2025).

WHO (2015b) *World Report on Ageing and Health*. World Health Organization, Geneva, Switzerland. Available at: https://apps.who.int/iris/handle/10665/186463 (accessed 5 November 2023).

WHO (2022) *Health Systems Resilience Toolkit: A WHO Global Public Health Good to Support Building and Strengthening of Sustainable Health Systems Resilience in Countries with Various Contexts*. World Health Organization, Geneva, Switzerland.

WHO (2024) *City Leadership for Age-Friendly Communities in the Post-Pandemic Era: Five Lessons for Building Health Emergency Resilience from 16 European Cities*. WHO Regional Office for Europe, Copenhagen.

WHO and EU (2017) *Age-Friendly Environments in Europe: A Handbook of Domains for Policy Action*. World Health Organization Regional Office for Europe, Copenhagen, and European Commission, Brussels.

# 8 Sustaining Urban Food Gardens for Community Resilience: A Case Study of the Langa Agrihub, Cape Town, South Africa

**Kurt Ackermann[1]* and Tamsin Faragher[2]**

*[1]South African Urban Food and Farming Trust (SAUFFT), Cape Town, South Africa; Global Risk Governance Programme, University of Cape Town, South Africa; [2]City of Cape Town, Risk and Resilience Department, Cape Town, South Africa; African Centre for Cities, University of Cape Town, South Africa*

**Abstract**

In an era marked by compounding risks and cascading systemic failures, community-level initiatives offer important insight into adaptive responses that build resilience from the ground up. This chapter contributes to the volume's focus on decision making and risk mitigation under climate uncertainty through a case study of the Langa Agrihub in Cape Town, South Africa. Positioned within a township historically shaped by racialised spatial planning and ongoing socio-economic exclusion, the Agrihub supports a network of over 250 urban farmers comprised of more than 30 food gardens. It provides access to inputs, infrastructure, knowledge exchange, and market connections, but more importantly, serves as a platform for civic participation, social cohesion, and local agency. Although community food gardens are often promoted as food security solutions, this chapter shows that their primary value lies elsewhere: in enabling communities to mitigate and adapt to intersecting stresses including economic marginalisation, environmental degradation, and infrastructure shortcomings such as water scarcity and energy insecurity. Based on three years of participatory implementation, evaluation, and reflection, the chapter documents the Agrihub's contributions to economic livelihoods, mental and physical wellbeing, biodiversity, climate adaptation, and democratic engagement. The Langa Agrihub illustrates how decentralised, place-based interventions can generate distributed resilience and complement formal risk governance systems. It challenges traditional, technocratic approaches to adaptation by foregrounding social infrastructure, informal networks, and lived experience as central to effective decision making. In contexts of high uncertainty and constrained institutional capacity, such models offer an important corrective to top-down planning paradigms. The chapter argues that urban agriculture, when supported through embedded community infrastructure like Agrihubs, should be recognised not only as a green or food policy tool, but as a strategic mechanism for strengthening community agency and resilience in the face of climate crisis.

## Introduction

Cape Town (South Africa), like other global cities, faces a severe and diverse set of shocks and stresses that create uncertainty and dynamic environments for urban actors (UN Habitat, 2022). Its resilience and vulnerability are influenced by chronic stresses triggered by poverty

*Corresponding author: kurt@fairfood.org.za

 *Climate Change and Risk Mitigation: Reducing Vulnerabilities and Enhancing Resilience* (ed. P. Barnes)
DOI: 10.1079/9781800624153.0008

and extreme inequality (McDonald, 2008; World Bank, 2018) at a national and local level.

These issues, together with poor economic growth and resultant unemployment, limit food access. Inflation and cost of living volatility from oil price increases, global conflicts, climate change (drought and flooding) and infrastructure shortfalls (e.g. electricity outages) affect food affordability. Conflicting household spending priorities pit food against others such as school fees, transport and energy, further influencing food access. Perversely, even though South Africa and the Western Cape grow surplus food for export, food insecurity is a pernicious risk to Capetonians (City of Cape Town, 2019). On the one hand, food is available, but on the other it is unaffordable and therefore inaccessible. As an underlying stress, food insecurity creates high vulnerability and links resilience to many other shocks and stresses that will be tested in the future by greater frequency and intensity – and by polycrisis events. For these reasons, food insecurity is a key measurement of resilience and a focus area for local government intervention.

Food gardens are widely promoted across stakeholders as an intervention to promote food security, in addition to economic opportunities. In practice, however, the South African Urban Food and Farming Trust (SAUFFT) has found that urban food gardens do not make significant direct contributions to addressing food insecurity. Still, food gardens can be valuable as supplemental strategies for food growers and their households to help mitigate some of the impact. This is particularly evident when food gardens within a community are supported collectively by an Agrihub.

Further, through the social networks of food growers established and strengthened by an Agrihub, urban food gardens also help meet a range of resilience objectives at a community scale and beyond, including mitigating climate risk associated with resource-intensive commercial production by producing food locally, conserving resources through social innovations that promote sustainable consumption and production, catalysing climate and environmentally sensitive approaches (Ley, 2019) such as water-sensitive development, activating public spaces to build social cohesion and networks, facilitating participation and engagement in governance processes and strengthening social infrastructure.

The Langa Agrihub case study illustrates ways that these objectives have been met, particularly in relation to building community resilience to climate change in a way that contributes towards preparing cities for an ever-changing and unpredictable future. Further, it has demonstrated the value of food growing as part of social infrastructure, an outcome strongly supported during crisis response to the shock of the COVID-19 pandemic.

Whether 'resilience' is perceived as ecological or urban – or both – social infrastructure is an important factor, with a municipality's ability to weather a disaster found to be dependent upon the size of social networks in its neighbourhoods and on the interconnection between the social networks (Resilient Cities Network, 2023).

## Context

Cape Town, South Africa's oldest and second-largest city, is located on the coast in the Western Cape Province. The city's population of 4.67 million (City of Cape Town, 2022) live in a 2461 km$^2$ area. Well known as a top international tourism destination, it boasts two UNESCO World Heritage sites and is responsible for 9.6% of national economic output.

Cape Town's resilience is affected and influenced by socio-economic challenges. It's economy, largely driven by finance, is also strongly influenced by agriculture. In fact, in 2021, the Western Cape agriculture, forestry and fishing sector grew by 8.9%, making it the largest contributor to the city's broader economic growth (Western Cape Government, 2022).

Despite sector growth, socio-economic issues persist. The City of Cape Town (the City) estimates unemployment at between 26% and 35% (City of Cape Town, 2023). National government further estimates that 45.9% of Capetonians live in poverty (COGTA, 2020). This is significant because of the linkage between unemployment and poverty (Ngubane *et al.*, 2023) and that between poverty and food insecurity (Battersby, 2011a). Despite 30 years of democracy, all three continue to manifest spatially in South African cities.

A 2021 national food security survey found that food inflation and rising living costs also negatively impact food access, and agency. It found that 15% of South African households have inadequate food access and 6% have severely inadequate access. The bulk of these households are in Johannesburg and in Cape Town, where 12.4% of households with children aged 5 years or younger reported experiencing hunger (StatsSA, 2023).

It follows that socio-economic factors underpin vulnerability and influence resilience to shocks and stresses, including drought, flooding and pandemics.

The 2015–2018 Western Cape drought, referred to as 'Day Zero', was caused by low rainfall across all Western Cape Water Supply System (WCWSS) catchments. Water availability for households and businesses, including food growers and producers, was severely constrained. Small businesses and poorer households who could not afford alternative water options such as boreholes were most negatively affected.

The 2020–2022 COVID-19 pandemic shone a bright light on food insecurity and the role of the informal food economy – including community kitchens, hawkers, spaza shops (informal, micro-convenience shops) and food gardens. Pre-pandemic, 80% of Cape Town's low-income households used the informal food economy to access food (Battersby, 2011b). Initial lockdown regulations in South Africa prohibited informal food trade (Skinner and Haysom, 2020). This prevented food access for households and curtailed traders' income causing food insecurity. Despite regulation changes that permitted food trading (and benefitted traders), food insecurity became a major crisis.

The scale and type of humanitarian response during the pandemic was unparalleled compared to other major shocks. Non-governmental organisations (NGOs) and community networks (including farmers) embedded in communities became critical for finding funding and providing humanitarian relief in partnership with governments, NGOs, community groups and international donors (South African Food and Farming Trust, 2020). At this time, the contribution of farmers was elevated, noting that 'farmers' also refers to 'growers', food growers', 'food gardeners' and 'market gardeners'.

During 2023, flooding caused significant crop and infrastructure damage in rural areas and to homes and livelihoods in urban areas, including Cape Town. Flooding is, however, not unusual in Cape Town. Poor spatial planning and urban stormwater management combine with natural groundwater fluctuations causing flooding and flood exposure. Informal settlements and farmers are often located in flood risk areas, where they also experience extreme heat days and heatwaves (de Wit *et al.*, 2023).

Governance shortcomings in policy, planning, strategy and accountability emerged during the management of these climate-related shocks and are apparent in other basic service sectors, such as electricity. Poor governance allowed the 'capture' of South Africa's national electricity utility in 2008. This enabled high levels of corruption and theft, affecting maintenance and infrastructure development that left the utility unable to provide sufficient electricity to meet the country's full demand (Daniels, 2023).

Supply was managed using planned outages, known locally as load-shedding. Load-shedding escalated and, in 2023, Capetonians experienced it on 335 days. Estimates in 2022 suggest resultant gross domestic product (GDP) contraction of between 2 and 3% (World Bank, 2023). In addition to job losses and reduced income, load-shedding and the prohibitive cost of energy alternatives has changed household food-related choices towards foods that require little or no cooking or refrigeration. These foods are often unhealthy, processed or ultra-processed (Haysom and Pulker, 2023).

As demonstrated, shocks and stresses significantly affect socio-economic factors that create new and exacerbate existing vulnerability to climate and infrastructure-related shocks and stresses – including food insecurity.

The City's policy response to food insecurity is twofold. On the one hand, it is addressed in the Resilience Strategy (2019) (the Strategy; City of Cape Town, 2019), and on the other, via the Climate Change Strategy (City of Cape Town, 2021a) and Climate Change Action Plan (City of Cape Town, 2021b).

The consolidation of food systems and climate change within one department, the Risk and Resilience Department, makes possible the important shift from understanding urban agriculture as a strategy for promoting food security (which is ineffective) to understanding it as a strategy for climate change mitigation and adaptation, which the UN considers 'a necessary response' and key strategy (FAO *et al.*, 2020).

As per the Resilience Strategy, a Food Systems Programme (the Programme) was developed using a food systems approach. It takes cognisance of these research findings: (i) food gardens do not produce enough food for household food security, and (ii) they do not generate enough food-growing income to transition out of poverty. These findings are contradictory to the City's Urban Agriculture Policy (City of Cape Town, 2014) and the Food Gardens Policy in Support of Poverty Alleviation and Reduction (City of Cape Town, 2013). These policies are not embedded within a City department and consequently have not been implemented for some time. Other policies, such as the Green Infrastructure Programme, consider food gardens more broadly as part of a 'green' network. The latter aligns with the Programme's approach that recognises food gardens' value in providing dietary diversity and fresh produce, in addition to heat adaptation, groundwater management, social cohesion and social infrastructure for community resilience.

Non-government actors, including civil society organisations and NGOs such as SAUFFT, play important roles supporting government. SAUFFT is based in Cape Town and works both independently and in partnership with the City to strengthen urban communities' resilience through food and farming.

SAUFFT's journey commenced in 2012 with the establishment and management of the Oranjezicht City Farm, a non-profit community food garden (South African Food and Farming Trust, 2019). The farm demonstrates the green infrastructure potential of food gardens and is a model of how community food gardens can positively impact the lives of their farmers and the surrounding community (International Civil Society Centre, 2020) including through social cohesion, human presence in places, healthy natural systems, collective agency, nutrition and food security.

SAUFFT and other NGOs have stepped into the City's food gardens policy and organisational vacuum. Success includes training 5000 farmers in poor communities, with hundreds of community food gardens (Paganini *et al.*, 2019) evincing interest through ongoing activity. While these food gardens have the potential to contribute to the resilience of their communities, there is a high attrition rate, particularly in under-resourced and vulnerable communities like Langa, the study area for this chapter. A food garden's range of influence is further limited to its immediate neighbourhood and does not reach a community scale.

Provincial strategic work in 2017 first focused on rural smallholder farmer-focused Agrihubs (Southern Africa Food Lab, 2017) and secondly explored the relevance of Agrihubs for their urban, informal food garden context (South African Food and Farming Trust, 2019). The emergent Agrihub concept addressed limiting issues by providing an infrastructure-based facility, with a farmer-centred, farmer-led cooperative network for innovation.

The outcome of the exploration was the separation of the roles of Agrihubs and food hubs into different but supporting roles. An Agrihub primarily supports the 'farming' or production aspects of farming, whereas food hubs primarily support the 'marketing' or sales and distribution aspects (Table 8.1).

The Agrihub approach (Table 8.2) has five objectives and promotes a farmer-centred approach based on farmer agency. It leverages off farmer collaboration and cohesion, forming a focal point for their needs within a geographic area.

By supporting urban farmers, particularly non-commercial or informal farmers, Agrihubs help with household and community resilience to shocks and stresses. The farmers participating in the Agrihub form a network that provides benefits beyond the more transactional functions supporting food production. This social infrastructure at a community scale plays a pivotal role in building multi-dimensional resilience and providing support during crises. However, government does not recognise the role of food gardens as social infrastructure, or their contributions to resilience and crisis response. Agrihubs play a potentially catalytic role in unlocking the value of food gardening

**Table 8.1.** Functions of an Agrihub compared to a food hub. Based on Southern Africa Food Lab, 2017.

| Agri-food hub function | | | | | | | |
|---|---|---|---|---|---|---|---|
| Agrihub | | | | Food hub | | | |
| Foster knowledge generation & innovation | Input services | Value addition for farmers | Coordination: Market agent function | Coordination: Aggregation, distribution & marketing | Better market partnerships | Assistance to meet buyer requirements | Branding & marketing strategies |
| Share best local practices in farming | Nursery, Extension support, Tool hire, Farming inputs | Sorting facilities, Pack houses, Processing infrastructure | Availability and order fulfilment to assist with market access | Locally produced foods from multiple producers to multiple markets | Working with producers as business partners instead of suppliers | Providing technical assistance or finding partners to provide it | Identity preservation, group branding, specialty product attributes, certifications, price assurance for producers |

**Table 8.2.** Agrihub objectives. Author's own table.

| Objective | Rationale |
|---|---|
| To create increased economic opportunity for local farmers in under-resourced communities | • Reliable, accessible, affordable inputs → lower cost, higher yield<br>• Improved knowledge and skills for food growing and access to further learning → improved efficiency and reduced risk/losses<br>• Improved market access → more predictable, higher income<br>• Reduced risk → increased investment → increased yield/ improved quality → increased income/margins |
| To create improved local access for community members to healthier food choices | • Shorter time from harvest, harvested riper (local transport only) → more nutritious<br>• Less waste risk → easier to keep in stock for local retailers and local restaurant use<br>• Local production → consumer knowledge and trust<br>• More community members growing their own food for personal/household consumption |
| To increase economic opportunity and improved livelihoods for community members and local businesses | • More resources (locally grown produce, organic waste)<br>• Market access (local) – knowledge, support, market needs, trust<br>• Recirculation of locally spent money<br>• Local livelihood opportunities in the food economy |
| To increase the agency of community members to shape their local food economy | • Choose locally produced food, bought locally<br>• Choose to 'import' food when necessary, from 'good' sources (regeneratively farmed, local/peri-urban/regional, direct from farmers or farmer cooperatives, etc.)<br>• Engage in entrepreneurial opportunities (informal and formal)<br>• Collective governance |
| To establish a sustainable, self-governing, independently operating Agrihub | • No dependency on outside funders and resources<br>• Members determine priorities → relevant and valued<br>• Local governance → greater accountability to local community |

by providing support and services, which aligns with and complements local government's efforts to improve resilience.

This case study will take a deeper look at the ways an Agrihub has done this in practice, based on three years of establishment and operations. The first Agrihub was launched in the community of Langa as a pilot to (1) support a network of existing community food gardens and (2) test emerging theories and assumptions as informed by experience from the Oranjezicht City Farm (and others) and from the pandemic. While three additional Agrihubs have since been established in Cape Town, the outcomes of the Langa pilot are central to determining the potential of Agrihubs to further build and support community resilience in Cape Town as well as inform future policy.

This chapter proceeds with the Langa Agrihub Case Study, followed by challenges, outcomes and business case, before concluding.

## Case Study: Langa Agrihub

The Langa Agrihub is in Langa, Cape Town's oldest formal township, which was established under the Native (Urban Areas) Act of 1923. Through the Act, black South Africans were segregated into designated locations. This isolated them from the broader economy and created labour reservoirs for menial work in white-owned industries and white households (Teagle and Sanchez, 2023). Today, Langa is still home to predominantly black African, Xhosa-speakers (StatsSA, 2013), most of

**Fig. 8.1.** Map of food gardens in Langa (green place markers) with location of Agrihub (red circle). Base image source Google Earth 2024.

whom maintain socio-cultural links to traditional Xhosa areas in the Eastern Cape province which remain predominantly rural villages with subsistence farming practised extensively.

Langa's residents face numerous persistent day-to-day challenges. These include low average monthly household income, estimated (January 2024) to be R2144, with 72% living below the poverty line and 51% of residents reporting that they have insufficient food to eat. Approximately 40% are unemployed and just 7% of residents have completed secondary education (Pulker, 2016). Across Langa, high levels of violence, crime and social unrest are experienced (Mzamo, 2013), with the effects on young residents in particular linked to high levels of protest and unrest in the community (Nleya *et al.*, 2011).

The Langa Agrihub (the Agrihub) is centrally located on Lerotholi Street (Fig. 8.1). The land is leased from the owner, the Red Cross Society, by the Masakhe Foundation, a non-profit local community development organisation. The lease is for 10 years and the leased portion measures 20 × 30 m. It forms part of a precinct that includes a community hall (separately managed by the Red Cross Society), food garden, art gallery and coffee shop.

In early 2019, SAUFFT began to work with a food garden in Langa. During this time, it expanded the food garden network by reaching out to other organisations supporting growing sites, as well as several independent food gardens. The pandemic, however, interrupted progress, but by March 2022, the network had 23 members (University of Cape Town, 2022). Through networks established during the pandemic, SAUFFT sourced international funding for the Agrihub implementation, alongside provincial government funding for infrastructure, enabling construction to commence as lockdown restrictions eased at the start of 2022. The success of the approach has included the selection of a SAUFFT team who understand the cultural and social context. It follows that the team are qualified farmers, mostly black African, and first-language Xhosa speakers.

## Farmer collective

A group of 107 farmers from 23 community food gardens located in Langa established the farmer collective. Active recruitment and promotion in the community was via a two-day Open Day series of meetings and site visits hosted by SAUFFT and the farmers. It included a display of photos and maps and information about how people could get involved, what the expected benefits would be and contact details. The original farmer cohort grew to 273 farmers from 34 farms by the end of November 2023, in response to the Open Day and other promotional activities. The current cohort represents over 1% of the households in Langa, marking a 48% increase in farms and a 155% increase in farmers.

Economic aspects were prioritised because farmers were already spending money on inputs, and the Agrihub could supply their needs more conveniently, cheaply and at higher quality (Chamberland, 2022). Beyond providing practical value to farmers, the Agrihub needed a sustainable business model not reliant on volunteering or donations. Knowledge exchanges, extension services, access to tools, access to market and other benefits enhanced the value to the farmers. The approach centred the agency on the farmers themselves (e.g. Battersby and Marshak, 2013; Paganini *et al.*, 2019), who drew from experience informed by participation in church groups, street committees, savings groups and other collective social structures, and resulted in the decision to form a steering committee. The steering committee, elected from collective members, is responsible for directing operations and governance.

With a farmer collective directing operations and controlling governance, the tension between Agrihub needs and individual farmer can be managed by the farmers themselves, reducing the risk of it being run as a for-profit business (including concerns raised in Tornaghi and Dehaene, 2020) or of SAUFFT becoming a gatekeeper. The initial appeal was financial and transactional around inputs, but the ultimate benefit comes from the combination of all the products and services, which, together, strengthen the food gardens' resilience.

## Establishment

In 2020, a food garden and Agrihub landscape masterplan and vision were developed with the farmers. Its purpose was to illustrate a collective vision that imagines a beautiful and functional space, while also guiding development and attracting potential investors.

The Agrihub layout (Fig. 8.2) draws from the original landscape masterplan. It is a U-shaped arrangement of small structures around a multipurpose central courtyard, including dry storage (A), a facilities control room (B), an office (C), cold storage (D), a kitchen (E) and an ablution block with shower (F). Initial infrastructure improvements provided a well point, pump, storage tanks and (temporary) chemical toilets. In 2023, the site was cleared, compacted and levelled and a perimeter fence and gates installed along with a security camera system.

During construction, activities were enabled using temporary infrastructure such as shade structures, tables and a refrigerated trailer. In this way, it was possible to host promotional 'pop-up' market days to build the customer base. Paying and non-paying customers included local restaurants, informal street traders, informal spaza shops, shops, caterers, feeding scheme coordinators, community kitchens and food retailers. Even though the infrastructure was incomplete during construction, it was still possible to host farmer training workshops and to receive and distribute bulk inputs.

## Current status

All buildings were externally complete in early 2024. The cold store and kitchen internal fit-outs were nearing completion, but all others were complete, allowing Agrihub activities to take place as planned. The current focus is on regularising day-to-day operations and completing the implementation of secondary infrastructure such as seedling tunnels and irrigation (Fig. 8.3).

Existing funding will sustain SAUFFT's support of the Agrihub's development for at least another 2 years. They will help the steering committee towards governance and management independence, help Agrihub operations become financially self-sustaining and strengthen the knowledge and skills within the farmer network.

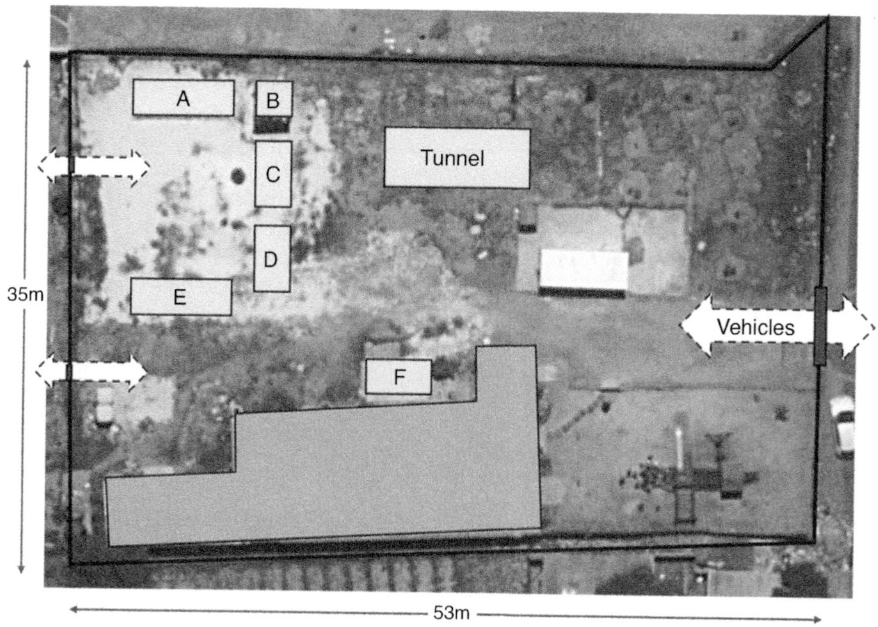


**Fig. 8.2.** Agrihub layout (2022 aerial photograph base layer). From Leslie, 2023.

**Fig. 8.3.** Photo of Langa Agrihub from the east (September 2024). Author's own image.

(Further details on the business model are provided below.)

In August 2023, SAUFFT commissioned an external evaluation of the Agrihub to assess if the implementation was meeting the needs of farmers, and to support evidence-based decision making to improve programme delivery. Focus groups from seven food gardens participated in the evaluation process, which includes surveys, interviews and focus group discussions (Leslie, 2023).

Key survey results:

- More than 87% of respondents claimed to be getting more money from their food garden.
- 87% of farmers ($n = 26$) perceived increased yields since receiving support.

- 93% of farmers ($n$ = 28) perceived that support had improved their food garden's performance.
- 78% of farmers who received products ($n$ = 21) said that Agrihub products were better than those supplied by other places.
- 60% of surveyed farmers ($n$ =18) stated that their food garden is performing well enough to meet their goals.
- Average farmer rating of the Langa Agrihub was 7.2 out of 10 (SD ± 2, $n$ = 30).
- 100% of respondents said that they would recommend food gardens to other urban farmers.

The following two sections – challenges and outcomes – articulate key points from the evaluation report.

## Challenges

Challenges are contextualised within the broader context of shocks and stresses experienced by Cape Town residents, as well as those specific to Langa. These are over and above the challenges of being a farmer in a vulnerable, under-resourced community.

Farmers identified dumping, relationships, communication with Langa Agrihub, insufficient extension training, poor soil quality, lack of tools, cultivation and water challenges, lack of soil inputs, infrastructure challenges, theft in food gardens, lack of seedlings, lack of space and market access as challenges (Fig. 8.4).

While the extent of challenges varied across food gardens, all respondents believed that the Agrihub could support their food gardens better (Leslie, 2023). Agrihub operation and communication across the network of farmers, particularly between farmers and the steering committee, were specific areas identified for improvement.

The external evaluation was supported by internal evaluation by the SAUFFT team, who practise reflective learning by capturing challenges and lessons learned in learning briefs for quarterly project reports. These include challenges not uncommon to NGO development work in vulnerable communities and those associated with working at nature's pace according to seasonal and annual cycles.

## Outcomes

This section discusses outcomes thematically and draws on the evaluation report as well as observations and data collected by SAUFFT fieldworkers through formal and informal interviews with farmers, analyses by researchers, engagements with partners and other stakeholders, amongst others. SAUFFT and the Agrihub steering committee use these data to gain a diversity of insights and perspectives on how the project is effecting changes in the lives of the farmers and the community of Langa, as well as how the project is achieving its stated goals.

### Economic well-being

Levels of benefit derived from the Agrihub were measured at a farmer, household, food garden and community level.

> Ohh yes, [the Agrihub] definitely worked for me because I was just farming. So, the money was coming very slow. But now at the market [day], I gain a lot which helps me to even buy food at home, and even buy a T-shirt for myself. (Langa Agrihub farmer)

Farmers have accessed cheaper inputs and markets, and generated income (Leslie, 2023). SAUFFT fieldworkers estimate that the out-of-pocket costs of inputs (seeds, seedlings, compost and manure) are 67% lower on average. In addition, there is a cash return on inputs of approximately R10 for every R1 invested.

Despite these favourable returns, delays in planting seedlings cause them to perish if not planted within 2 weeks. Consequently, 51.2% of available seedlings are not harvested, reducing yields and contributing towards a 'harvest gap'. Pests, theft and poor cultivation methods are other issues that affect harvest success. It follows that addressing these issues could increase harvests by roughly half.

Even with the harvest gap, the 54% of farmers who sell their harvest through the Agrihub have seen the value of their harvest increase by 38.9% on average. The price received when sold through the Agrihub is 217% higher than if they sold independently to neighbours, local business or community

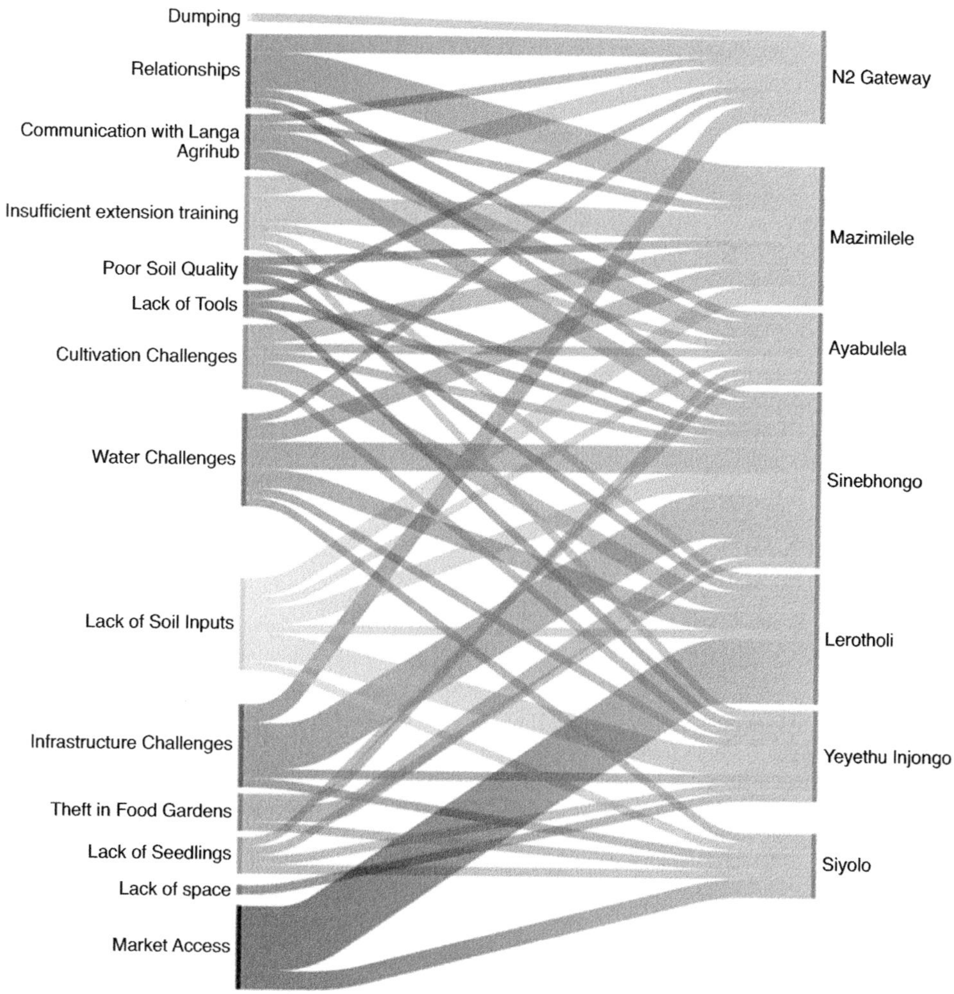


**Fig. 8.4.** Langa Agrihub challenges (left) identified by the farmers at each of the surveyed collective food gardens (right). Note: The width of the connecting lines indicates the amount of relevant information shared, not the significance of the information. From Leslie, 2023.

kitchens. The average harvest value per active farmer per month is R510, while the cash income is R119. The additional income represents 23.7% more to a farmer's household in a combination of savings on food expenses and cash income.

Other ways that harvests are directed are own consumption or household processing, donation to others, donation to community kitchens and school feeding schemes, and/or independent sales.

## Social well-being

Urrego Diaz *et al.* (2025) found that even though farming is a means to many ends, it is mainly a coping strategy or mechanism with social benefits related to local challenges such as food price increases on the one hand, and general issues such as the COVID-19 pandemic and climate change on the other. Social wellbeing is nurtured through Agrihub activities, including the steering committee, workshops and site visits. These activities promote connecting and sharing between farmers with common interests that may be related to successful production, identifying trustworthy neighbours or developing new collectively beneficial initiatives such as a seedling garden.

In this way, farmer social networks have been built around the Agrihub. Community cohesion and resilience is measured in part by knowing your neighbours' names (Arup, 2023). The social networks developed around the Agrihub contribute towards these objectives and provide coping mechanisms for local challenges, such as food access that is improved by enabling optimal farming through the supply of affordable inputs that results in affordable produce and thereby improves access. Optimised farming also increases production outputs and enables greater potential farmer support of community events, e.g. funerals, by contributing meals cooked from self-grown produce (Leslie, 2023).

Farmers who live a distance from their farms have by necessity become acquainted with the neighbours living adjacent to their farms to solicit their oversight, illustrating how the networks built to support food gardens have resulted in positive interactions and outcomes for farmers (Heming and Mgcoyi, 2022) that also promote social wellbeing. Farmers have also noted shifts in community perspectives on gender norms (Heming, 2022).

## Physical and mental health

Farmers recognise the diverse physical and mental health benefits associated with farming. A more nutritious diet with vegetables, and reconnecting with indigenous practices and herbal medicine (Urrego Diaz *et al.*, 2025) are a few that they have identified. Ablutions and showers at the Agrihub also help with physical hygiene while providing comfort and dignity.

Growing food makes farmers feel hopeful because of the potential income and improved health. The farmers have identified a wide range of other associated mental health benefits, from coping with isolation and stress (Urrego Diaz *et al.*, 2025), brain fog, fatigue and memory loss (Landsman, 2023) to managing depression and anxiety (Heming, 2022).

The Agrihub also has mental health benefits. Participation in steering committee meetings and planning discussions has been found to reduce farmers' sense of isolation, as has having their concerns heard and advice sought. Older farmers expressed hope for the future, emanating from the belief that the Agrihub will continue and attract young farmers (Leslie, 2023).

## Safety

Langa experiences high levels of violence, crime and social unrest (StatsSA, 2013) that cause personal and farm safety concerns. Incidents, typically theft, attempted theft or damage to farm property, usually occur at night when farmers are absent. Produce and seedling theft occurs, but is less of an issue than infrastructure and materials (electrical wires, pumps, metal tools, sections of fencing, etc.) theft.

The Agrihub, in contrast to most of the food gardens that have limited security, is secured with a 2.1 m high perimeter fence and gates that provide security and access control. A security system was also installed with cameras to monitor the perimeter and an alarm system for deterrence and activation of a security response. These were installed and a security company appointed after a major theft of building materials in a targeted incident. Farmers further mitigate the risk of theft by storing expensive tools in a secure tool 'library' where they can be 'checked out' and 'in', because even though theft from farms is uncommon, in a resource-scarce context, damage to small items, like an irrigation pipe fitting, can cause substantial problems.

Because daytime personal safety is less of an issue, farmers feel comfortable bringing children, grandchildren, friends and visitors to their farms without concern, including welcoming visits from school groups. When more structured events are organised, such as large groups of visitors, the local 'Langa Safety Forum' monitors are engaged. The forum provides security through their presence, which signals to the community that the visitors have taken care to come to Langa in a respectful way that also brings some form of benefit to them.

## Civic engagement

Langa Agrihub's network of more than 250 farmers recently linked with other Agrihubs that include hundreds more farmers. Through these interactions, many have started to understand themselves as being part of a collective, and that this can manifest in a collective voice for the farmers, which could speak, and even advocate, on their behalf. This shift in self-understanding is consistent with research indicating that the Agrihub has decreased barriers and increased opportunities for farmers, thus increasing their 'aspirational window' and, as a result, their capacity to aspire (Chamberland, 2022), and in contrast to a former reluctance to participate due to negative experiences with government, including local government.

The farmers' first exercise using their Agrihub-enabled voice was spurred by an opportunity to comment on the City's revised Urban Design Policy, which was available as part of the public participation process. SAUFFT mobilised the farmers and hosted a workshop to assist in understanding the policy process and the document (which was not available in their home language). Farmers' inputs and comments were gathered and consolidated into a document that was submitted to the City. Apparent through this process was the farmers' developing self-awareness within the regulatory and conceptual landscape that they could challenge key foundational concepts including the role of policy and of representation during the policy development process itself, even before it is circulated for public participation.

Growing interest, among some of the farmers at least, in engaging more actively in these previously inaccessible civic processes has extended to engagement in an upcoming Environmental Management Strategy review process.

As the other Agrihubs develop and link together in a larger collective, it may be that they become a centre for the voice for urban farmers in Cape Town that grows with confidence to speak to their own interests, both within official processes, and in contesting or subverting them. This outcome is consistent with Dehaene *et al.* (2016), who argue that urban agriculture can create localised seedbeds for an 'insurgent form of social change'.

## Urban green spaces

The Agrihub farms comprise a total area of 10,605 m$^2$. Of this, 4083 m$^2$ is under productive cultivation and 1032 m$^2$ is planted borders. The balance is working space, paths and non-productive land. The planted area of the farms comprises 0.35% of the total area of Langa. It is distributed throughout the community, providing tended pockets of green such that nearly all Langa residents are within 250 m of one or more. Since the establishment of the Agrihub, it is estimated that green space has increased by 2600 m$^2$, which is roughly 25%.

The increase in green space access is significant because of the positive affect green space has on human health and wellbeing (Hartig *et al.*, 2014). Increasing green spaces in vulnerable and under-resourced communities like Langa therefore has the potential to reduce the inequality of health outcomes (Allen and Balfour, 2014).

Further, activated urban green spaces provide 'eyes on the street' that improve community safety. They also secure properties for food growing, protecting them from illegal dumping, informal settlement and the loss of public or other open space needed for ecosystem services.

## Climate adaptation and mitigation

Both food gardens and Agrihubs, via localised food growing and supply, play roles in climate adaptation and mitigation. Adaptation benefits of food gardens include cooling; groundwater recharge and reduced flood risk; and soil biodiversity contributing towards mitigation. Agrihubs, on the other hand, can play a role in adaptation as cooling hubs, whereas mitigation can be achieved through lower carbon emissions as a result of shorter value chains and eliminating use of imported oil-based fertilizers.

City policy broadly promotes urban 'greening' and, in particular, the planting of trees for heat adaptation, as trees have a cooling effect via transpiration and irrigation that cools the earth around them. Food gardens usually have limited shade and seek to maximise sunlight for crop production. This is true for most of the Langa farms that are structured into beds for row crops and perennial herbs and enclosed with perimeter shrub windbreaks. Other factors that contribute to low tree-planting levels in food gardens include high wind exposure that stunts growth, the preference for annuals that have a shorter investment period and low tree cultivation knowledge. Productive trees have, however, been planted at the Agrihub for demonstration and for workshops to improve tree-growing knowledge and contribute towards increasing the urban forest.

While the City's Heat Action Plan promotes 'greening' on the one hand, it also promotes cooling centres, particularly for areas where trees are difficult to grow and there are high levels of social vulnerability. The concurrent availability at the Agrihub of a support network, skilled and motivated growers, water, shade and greenery make it well-suited for this function. The central location within Langa and the farmer collective further enhance this potential role. SAUFFT have initiated conversations with farmers and will be piloting a 'cooling station' at the Agrihub in 2025, with the potential to expand to other farm sites in the community.

## Biodiversity and soil biodiversity

The Agrihub actively promotes biodiversity through workshops and knowledge exchanges held across 25 gardens. In addition, 2200 medicinal and culinary plants from 10 species, six indigenous to the Cape Floristic Region, were planted at the Agrihub. The importance of biodiversity for food growing and beautification has resonated with the farmers. Of the 15 farms where biodiversity gardens were established, 12 continue to be maintained for this purpose, illustrating the normalisation of biodiversity within food gardens (South African Food and Farming Trust, 2022).

Soil biodiversity also plays an important role in addressing global climate change. It contributes to the reduction of greenhouse gas emissions and to absorbing carbon and fixing nitrogen from the atmosphere into soils. It is typically associated with the maintenance and enhancement of above-ground biodiversity, in addition to reducing threats to ecosystem services and bioremediating contaminated soils (FAO *et al.*, 2020). This is an important consideration in urban contexts where soils are frequently exposed to contaminants, such as Langa, where issues such as illegal dumping and sewer leaks are common.

## Environmental quality

Green spaces have the potential to improve soil and groundwater quality as well as stabilise sandy soil.

The soil at the farms in Langa have low levels of organic material and elevated salinity. This has been improved by removing rubble and other material, watering the soil to sustain and stimulate naturally occurring and dormant microbial biodiversity and adding compost, vermicompost, manure, mulch and wood chips to increase the organic matter in the soil.

Some of the added compost used in the Agrihub farms was sourced from a City compost pilot focused on determining optimal ways to divert food waste generated by fresh produce traders away from landfill to compost. The Agrihub provided training for the compost makers on behalf of the City. Other services provided included quality assurance and coordination and distribution to farmers (South African Food and Farming Trust, 2022).

Farmers in Langa commonly irrigate using groundwater from well points. The high water table (less than 30 m deep) makes groundwater relatively accessible, but it also exposes it to contamination. Farmers have a direct interest in water quality and availability and consequently play an important management role in reporting issues to the Agrihub or City, in addition to their farming activities that support groundwater recharge.

Another environmental hazard urban agriculture has been shown to mitigate is air pollution. Research indicates that air pollutants contribute to rising temperatures and heatwaves that cause heat stress and deaths from heart attacks or strokes (WHO, 2015), especially among elderly people and those with non-communicable diseases (Kenny *et al.*, 2010). Mitku *et al.* (2023) also found that exposure to particulate matter that is 2.5 μm or less in diameter and sulphur dioxide pollution adversely affects birth outcomes. Gardens improve air quality and thereby play a role in mitigating these risks.

## Community resilience

Urrego Diaz *et al.* (2025) found that the Agrihub has created a continuous platform to drive more systematic interaction for knowledge management and exchange that help farmers to cope with crises. For example, farmers were able to support one another during a week-long public transportation strike by minibus taxi operators. The strike was violent and included sporadic acts of vandalism and crime. Langa residents were unable to travel to work or school, or to access health care outside the community. The strike also disrupted logistics more broadly, including deliveries of food, medicine and other essentials (Hirsch, 2023). As scarcity of food and risks of violence increased, farmers exchanged information in person and through WhatsApp groups across the community. Through these communications, they were able to share food and help keep safe.

## Business Case

The financial sustainability of the Agrihub is predicated on the capital and establishment costs being paid by donor funding, and the ongoing operating costs being proportional to the level of usage by the farmers, meaning that the passive operating costs are minimised when not in use. This is accomplished through infrastructure design, such as electricity from solar panels, water from a well point, insulation and natural ventilation in structures, and more. Repair and maintenance costs are similarly minimised by requiring only materials, skills and tools that are widely available locally. When the site is used, the associated costs are known to the steering committee, which makes the decisions regarding how much money the participating farmers need to contribute, either in terms of mark-up on inputs sold to farmers, or in terms of a proportion of income from produce sales that is retained by the Agrihub before paying out the farmers. Farmers commit to supporting the financial sustainability of the Agrihub when they join.

The Langa Agrihub is one of four that have been established in an ecosystem where each has its own specialisation based on its local context and strengths, which simultaneously contributes to the others. The Agrihub in the community of Khayelitsha has emerged as the most commercially focused, and more productive and financially motivated farmers who may be based in Langa are choosing to supply their produce to the Khayelitsha Agrihub to access markets that are larger and more frequent buyers like restaurant chains, hotels, small food production businesses and others. The Langa Agrihub's markets, by comparison, are less frequent and more locally focused. However, Langa is seeing more tourist activity and school group educational visits, which also bring income.

The Agrihub approach is not prescriptive about the ways that income needs to be generated, avoiding a 'cookie-cutter' model and encouraging local differentiation that remains valuable to the ecosystem of Agrihubs.

## Conclusion

Food insecurity underpins poor communities' vulnerability to shocks and stresses – and undermines their resilience. This case study illustrates that while food gardens are not the solution to food security, they do play multidimensional roles that are enhanced by support from Agrihubs and thus play an important role in building community infrastructure and resilience.

## References

Allen, J. and Balfour, R. (2014) *Natural Solutions for Tackling Health Inequalities*. UCL Institute of Health Equity, London. Available at: https://www.instituteofhealthequity.org/resources-reports/natural-solutions-to-tackling-health-inequalities/natural-solutions-to-tackling-health-inequalities.pdf (accessed 15 July 2025).

Arup (2023) City Resilience Index. Available at: https://www.arup.com/insights/city-resilience-index/ (accessed 15 July 2025).

Battersby, J. (2011a) Urban food insecurity in Cape Town, South Africa: An alternative approach to food access. *Development South Africa* 28(4), 545–561.

Battersby, J. (2011b) *The State of Food Insecurity in Cape Town*, AFSUN Food Security Series. Idasa Publishing, Cape Town.

Battersby, J. and Marshak, M. (2013) Growing communities: Integrating the social and economic benefits of urban agriculture in Cape Town. *Urban Forum* 24(4), 447–461.

Chamberland, J. (2022) *Farmer Aspirations in the Langa AgriHub*. University of Cape Town directed study project report.

City of Cape Town (2013) *Food Gardens Policy in Support of Poverty Alleviation and Reduction*. Available at: https://resource.capetown.gov.za/documentcentre/Documents/Bylaws%20and%20policies/Policy_Food_Gardens.pdf (accessed 29 July 2025).

City of Cape Town (2014) Urban Agriculture Policy. Unpublished.

City of Cape Town (2019) Cape Town Resilience Strategy. Cape Town. Available at: resource.capetown.gov.za/documentcentre/Documents/City strategies, plans and frameworks/Resilience_Strategy.pdf (accessed 15 July 2025).

City of Cape Town (2021a) Climate Change Strategy. Cape Town. Available at: https://resource.capetown.gov.za/documentcentre/Documents/City%20strategies,%20plans%20and%20frameworks/Climate_Change_Strategy.pdf (accessed 15 July 2025).

City of Cape Town (2021b) Climate Change Action Plan. Cape Town. Available at: resource.capetown.gov.za/documentcentre/Documents/City strategies, plans and frameworks/CCT_Climate_Change_Action_Plan.pdf (accessed 15 July 2025).

City of Cape Town (2022) Integrated Development Plan (summary). Cape Town. Available at: https://resource.capetown.gov.za/documentcentre/Documents/City%20strategies%2C%20plans%20and%20frameworks/IDP_2022-2027_Summary.pdf (accessed 15 July 2025).

City of Cape Town (2023) *Economic Performance Indicators for Cape Town*. Available at: https://www.investcapetown.com/wp-content/uploads/2024/08/41262-CCT-Epic-10-Year-Anniversary-Edition_interactive.pdf (accessed 29 July 2025).

COGTA (2020) Profile: Sekhukhune District 2 Profile: Sekhukhune District 3. Available at: https://www.cogta.gov.za/ddm/wp-content/uploads/2020/07/Take4_2020.06.25-SEKHUKHUNE-District-Profiles-Final-Version-.pdf (accessed 15 July 2025).

Daniels, N. (2023) 2023 cumulative load shedding worst in 15 years. *Cape Times*. Available at: https://iol.co.za/capetimes/news/2023-12-28-2023-cumulative-load-shedding-worst-in-15-years/ (accessed 15 July 2025).

Dehaene, M., Tornaghi, C. and Sage, C. (2016) Mending the metabolic rift – placing the 'urban' in urban agriculture. In: Lohrberg, F., Scazzosi, L., Licka, L. and Timpe, A. (eds) *Urban Agriculture Europe*.

de Wit, M., Rawlins, J. and Petrie, B. (2023) Economic risk assessment of climate change at the city level. The case of Cape Town, South Africa. *International Journal of Urban Sustainable Development* 15, 118–140. DOI: 10.1080/19463138.2023.2193813.

FAO, ITPS, GSBI, SCBD, EC (2020) *State of Knowledge of Soil Biodiversity - Status, Challenges and Potentialities*. FAO, Rome, pp. 174–177.

Hartig, T., Mitchell, R., de Vries, S. and Frumkin, H. (2014) Nature and health. *Annual Review of Public Health* 35, 207–228.

Haysom, G. and Pulker, A. (2023) *Scoping Review of Food Systems and Food Environments in Kisumu, Mbale, Tunis, Ouagadougou and Cape Town*. AfriFOODlinks. Available at: https://drive.google.com/file/d/1fFUmCS7FQglsrQWcSRq6SPlqUlqNUYM7/view?usp=sharing (accessed 29 July 2025).

Heming, S. (2022) Farmer Podcast: Nomandla Mthiyane. South African Urban Food and Farming Trust. Available at: https://fairfood.org.za/2022/03/04/farmer-podcast-nomandla/ (accessed 15 July 2025).

Heming, S. and Mgcoyi, C. (2022) Langa Farmer Profile: Sindiswa Lugulwana. South African Urban Food and Farming Trust. Available at: https://fairfood.org.za/2022/02/10/langa-farmer-profile-sindiswa/ (accessed 15 July 2025).

Hirsch, M. (2023) Taxi Strike Cost Western Cape R5-billion, MEC Tells Parliament. GroundUp. Available at: https://www.groundup.org.za/article/taxi-strike-cost-western-cape-r5-billion-med-tells-parliament/ (accessed 2 September 2024).

International Civil Society Centre (2020) Urban social farming practices and principles. In: *Civil Society Innovation and Urban Inclusion*. International Civil Society Centre, pp. 34–35. Available at: https:

//icscentre.org/wp-content/uploads/2021/02/ICSC_InnovationReport_2020.pdf (accessed 15 July 2025).

Kenny, G.P., Yardley, J., Brown, C., Sigal, R.J. and Jay, O. (2010) Heat stress in older individuals and patients with common chronic diseases. *Canadian Medical Association Journal* 182(10), 1053–1060. DOI: 10.1503/cmaj.081050.

Landsman, S. (2023) Langa Farmer Profile: Thobeka Gacula. South African Urban Food and Farming Trust. Available at: https://fairfood.org.za/2023/01/09/langa-farmer-profile-thobeka-gacula/ (accessed 15 July 2025).

Leslie, T. (2023) *A Process Evaluation of the Langa Agri/Food Hub Programme for the South African Urban Food and Farming Trust*. South African Urban Food and Farming Trust, Cape Town.

Ley, A. (2019) Community resilience and placemaking through translocal networking. Learning from Thailand and the Philippines. *The Journal of Public Space* 4(2), 165–178. DOI: 10.32891/jps.v4i2.1208.

McDonald, D. (2008) *World City Syndrome: Neoliberalism and Inequality in Cape Town*. Routledge, New York.

Mitku, A.A., Zewotir, T., North, D., Jeena, P., Asharam, K. *et al.* (2023) Impact of ambient air pollution exposure during pregnancy on adverse birth outcomes: Generalized structural equation modeling approach. *BMC Public Health* 23, 45. DOI: 10.1186/s12889-022-14971-3.

Mzamo, N.A. (2013) Assessing contributions of tourism development to poverty alleviation programmes in Langa, Western Cape Province. Doctoral dissertation, Cape Peninsula University of Technology, Cape Town.

Ngubane, M.Z., Mndebele, S. and Kaseeram, I. (2023) Economic growth, unemployment and poverty: Linear and non-linear evidence from South Africa. *Heliyon* 9(10), e20267. DOI: 10.1016/j.heliyon.2023.e20267.

Nleya, N., Thompson, L., Tapscott, C., Piper, L. and Esau, M. (2011) Reconsidering the origins of protest in South Africa: Some lessons from Cape Town and Pietermaritzburg. *Africanus* 41(1), 14–29.

Paganini, N., Karriem, R., Khan, Z., Kanosvamhira, T., Mfaku, A. *et al.* (2019) Rethinking required – how can urban agriculture in Cape Town still become sustainable in the future food system? Policy Recommendations and Results of the UFISAMO project.

Pulker, A. (2016) The relationship between urban food security, supermarket expansion and urban planning and policy in the city of Cape Town: A case of the Langa Junction mini mall. Masters dissertation, University of Cape Town.

Resilient Cities Network (2023) Resilience Hubs; RCities Peer-to-peer learning. Available at: https://resilientcitiesnetwork.org/downloadable_resources/Other/2023%20RCities%20Chief%20Resilience%20Officer%20Survey%20Results.pdf (accessed 29 July 2025).

Skinner, C. and Haysom, G. (2020) Feeding cities: Informal retailers play crucial role in urban food security. In: *Street Vendors and Public Space: Essential Insights on Key Trends and Solutions*. WIEGO, Manchester, UK. Available at: https://www.wiego.org/wp-content/uploads/2020/02/Street%20Vendors%20and%20Public%20Space%20-%20An%20EBook.pdf (accessed 15 July 2025).

South African Food and Farming Trust (2020) *Food Dialogues Report-2020*. Available at: https://foodsecurity.ac.za/wp-content/uploads/2020/11/Food-Dialogues-Report-2020-LR.pdf (accessed 15 July 2025).

South African Food and Farming Trust (2022) *EESI Project Report 2021-22*. Available at: https://law.uct.ac.za/sites/default/files/content_migration/law_uct_ac_za/1261/files/SAUFFT_report.pdf (accessed 29 July 2025).

South African Food and Farming Trust (2019) Oranjezicht City Farm. Available at: https://fairfood.org.za/oranjezicht-city-farm/ (accessed 29 July 2025).

Southern Africa Food Lab (2017) Workshop Report: Designing a Smallholder Farmer-Focused Agri-hub. Available at: https://www.southernafricafoodlab.org/wp-content/uploads/2017/03/Report_WCSmallholderFarmerHub.pdf (accessed 29 July 2025).

StatsSA (2013) *City of Cape Town 2011 Census Suburb Langa*. Statistics South Africa. Available at: https://resource.capetown.gov.za/documentcentre/Documents/Maps%20and%20statistics/2011_Census_CT_Suburb_Langa_Profile.pdf (accessed 15 July 2025).

StatsSA (2023) *Assessing Food Inadequacy and Hunger in South Africa in 2021 Using the General Household Survey (GHS)*. Statistics South Africa. Available at: https://www.statssa.gov.za/publications/03-00-20/03-00-202021.pdf (accessed 15 July 2025).

Teagle, A.S.D. and Sanchez, D. (2023) *Voices of Langa: A 100-year Social History of Cape Town's Oldest Township*. Human Sciences Research Council. Available at: https://hsrc.ac.za/news/inclusive-development/voices-of-langa-a-100-year-social-history-of-cape-towns-oldest-township/ (accessed 15 July 2025).

Tornaghi, C. and Dehaene, M. (2020) The prefigurative power of urban political agroecology: Rethinking the urbanisms of agroecological transitions for food system transformation. *Agroecology and Sustainable Food Systems* 44(5), 594–610.

UN Habitat (2022) The diversity of cities and visions for urban futures. In: *World Cities Report 2022: Envisaging the Future of Cities*, Chapter 1. Available at: https://unhabitat.org/sites/default/files/2022/07/chapter_1_wcr_2022.pdf (accessed 15 July 2025).

University of Cape Town (2022) Enhancing Impacts of Agricultural focused Green Entrepreneurs. Available at: https://law.uct.ac.za/grgp/outreach-eesi/green-entrepreneurs (accessed 15 July 2025).

Urrego Diaz, N., Ley, A., Ackermann, K. and Fischer, L. (2025) The Langa agri-food hub in Cape Town, South Africa: Strengthening farmers' networks and transforming food systems in crisis. In: Kingsley, J. and Egerer, M. (eds) *Crisis Gardening: A Global Perspective*. CRC Press, Boca Raton, Florida.

Western Cape Government (2022) *City of Cape Town 2020 Socio-Economic Profile*. Available at: https://www.westerncape.gov.za/provincial-treasury/sites/provincial-treasury.westerncape.gov.za/files/atoms/files/SEP-LG%202020%20-%20City%20of%20Cape%20Town.pdf (accessed 15 July 2025).

WHO (2015) *Reducing Global Health Risks Through Mitigation of Short-Lived Climate Pollutants. Scoping Report for Policymakers*. World Health Organization, Geneva, Switzerland. Available at: https://iris.who.int/bitstream/handle/10665/189524/9789241565080_eng.pdf?sequence=1 (accessed 15 July 2025).

World Bank (2018) *An Assessment of Drivers, Constraints and Opportunities Overcoming Poverty and Inequality in South Africa*. World Bank, Washington DC. Available at: https://documents1.worldbank.org/curated/en/530481521735906534/pdf/Overcoming-Poverty-and-Inequality-in-South-Africa-An-Assessment-of-Drivers-Constraints-and-Opportunities.pdf (accessed 20 April 2024).

World Bank (2023) South Africa: World Bank Backs Reforms to Advance Energy Security and Low Carbon Transition. World Bank. Available at: https://www.worldbank.org/en/news/press-release/2023/10/25/south-africa-afe-world-bank-backs-reforms-to-advance-energy-security-and-low-carbon-transition (accessed 15 July 2025).

# 9 Adaptive Urban Resilience Using Artificial Intelligence

**Haripriya Kesavan[1]*, Samhita R.[2] and Anshu Sharma[3]**
[1]*Spatial Planning and Product Technical Specialist, Resilience AI, New Delhi, India;*
[2]*CEO and Co-Founder, Innovation and Advocacy, Resilience AI, New Delhi, India;*
[3]*Co-Founder, SEEDS, New Delhi, India*

**Abstract**

(Disaster) risk is conventionally seen as an aggregated value for a population, usually at a settlement level. Early warning for an impending disaster triggered by natural hazards or weather variability is accordingly presented as a single interpretation of risk across a population. A heatwave is an extreme climatic event where unusually high temperatures make the environment dangerously hot, threatening people and livelihoods. However, every hazard will uniquely impact local communities and family groups, depending on a family's specific physical, social and economic characteristics. Artificial intelligence (AI) presents capabilities at very high efficiency levels to interpret hazard data and projected risk levels, allowing a rendering of projected impacts at a hyperlocal resolution, thereby enabling customised alerts and actionable insights for both short- and long-term risk reduction. This chapter examines the implementation and use of this technology, through the illustrative example of heatwaves in Delhi, India, and how it can support future benefits of risk and vulnerability reduction in local communities.

## Introduction

(Disaster) risk is conventionally seen as an aggregated value for a population, usually at a settlement level. Early warning for an impending disaster triggered by natural hazards or weather variability is accordingly presented as a single interpretation of risk across a population. A heatwave is an extreme climatic event where unusually high temperatures make the environment dangerously hot, threatening people and livelihoods. However, every hazard will uniquely impact local communities and family groups, depending on a family's specific physical, social and economic characteristics. Artificial intelligence (AI) presents capabilities at very high efficiency levels to interpret hazard data and projected risk levels, allowing a rendering of projected impacts at a hyperlocal resolution, thereby enabling customised alerts and actionable insights for both short- and long-term risk reduction. This chapter examines the implementation and use of this technology, through the illustrative example of heatwaves in Delhi, India, and how it can support future benefits of risk and vulnerability reduction in local communities.

## Do You Ever Feel Like Your Room Gets Super-hot in the Summer?

India constantly experiences some of the world's highest maximum temperatures, at an annual average maximum of 36°C and an

*Corresponding author: haripriya@resilience360.ai

 *Climate Change and Risk Mitigation: Reducing Vulnerabilities and Enhancing Resilience* (ed. P. Barnes)
DOI: 10.1079/9781800624153.0009

average monthly maximum of around 30°C (Word Bank, 2021). The country witnessed extreme weather events for a staggering 86% of days in the year 2023 (Centre for Science and Environment, 2023). Heatwaves, in particular, are becoming more frequent and prolonged, with a 2.5 day increase in total heatwave duration observed within established heatwave regions over the past 30 years. According to the Intergovernmental Panel on Climate Change (IPCC), a further escalation is expected, with an additional two heatwaves (12–18 days) impacting India by 2060, potentially even reaching previously unaffected regions in southern India. While in the past the focus of heatwave risk reduction was around outdoor exposure, the risk of indoor heat has been emerging as a significant threat.

Heatwaves, characterised by extended periods of abnormally high temperatures exceeding normal regional temperature variations, pose a growing threat due to climate change (Indian Meteorological Department, n.d.). These events, with their potential to make outdoor activities hazardous, expose a rapidly increasing global population to potentially life-threatening situations (World Health Organization, 2024). Heatwaves or hot weather conditions, despite their significant societal impact, including a rise in heat-related fatalities, often receive less attention compared to other natural disasters due to the gradual nature of their effects as they are not immediately as obvious as other climate hazards like cyclones (World Health Organization, 2020), which is why they are often referred to as 'silent disasters'.

In India, a heatwave is declared when the maximum temperature of a station reaches at least 40°C for plains and at least 30°C for hilly regions, as measured by the India Meteorological Department (IMD). Assessing heat stress on the human body requires a more nuanced approach than simply measuring air temperature. The Wet Bulb Globe Temperature (WBGT) is an indicator for assessing heat-related stress on humans, incorporating temperature, humidity, wind speed, sun angle and cloud cover (National Ocean and Atmospheric Administration, n.d.). This approach contrasts with the Heat Index (Indian Meteorological Department, n.d.), which considers temperature and factors in the humidity levels to estimate perceived human temperature. However, the lack of a globally standardised heat stress index necessitates the use of region-specific criteria for accurate risk assessment.

Thermometers alone are insufficient to capture the full spectrum of heat-related health risks, where risks associated with certain regions and people of varied socio-economic backgrounds are often underestimated (see Fig. 9.1). Effective mitigation strategies necessitate more sophisticated weather monitoring that incorporates regional variations in climate and human acclimatisation. According to Ravindra *et al.* (2024), solely relying on certain temperature thresholds fails to account for the significant spatial heterogeneity across India.

**Fig. 9.1.** Built environment and heat risk: illustration of how densely populated areas experience amplified heatwave effects. Content authors' own. Illustration by Arundhati Hakhu.

This approach overlooks the diverse climatic conditions and the varying levels of human adaptation to heat stress present in different regions. A more comprehensive understanding of these factors is crucial for safeguarding populations from the dangers of extreme heat.

In India, heatwaves predominantly occur from March to June, occasionally extending into July, with May witnessing the peak intensity. Notably, the Core Heatwave Zone (CHZ), encompassing states like Delhi, experiences the brunt of these events, characterised by the highest frequency and intensity of both regular and severe heatwaves, particularly during May.

A recent analysis by the World Weather Attribution initiative (2024) underscores the significant influence of climate change on India's extreme temperatures. The study reveals that climate change has increased by 45 times (Coleman, 2024) the likelihood of experiencing the current heatwave situation the country is facing. This translates to a 10% annual chance of India facing a severe late spring heatwave, highlighting the growing threat posed by a warming climate.

The climate resilience gap is the degree of unpreparedness of a nation or community. The bond between climate adaptation and climate mitigation is critical to reducing the climate resilience gap. Climate adaptation anticipates avoidable climate-induced losses by determining the climate risks, planning to reduce or mitigate risks, actioning prioritised interventions and scaling adoption. SEEDS (Sustainable Environment and Ecological Development Society), a national organisation working on community resilience, partnered with the AI for Good Lab of Microsoft to develop Sunny Lives, an open source AI model for hyperlocal risk assessment for vulnerable communities. The model was subsequently developed further into ResSolv, a version with applications to customised use cases of critical infrastructure, supply chains and large settlements at scale, hosted by Resilience AI, a climate-technology start-up. The initiative works towards determining climate risks with the speed and accuracy that can meet the urgency demanded by Sustainable Development Goal (SDG) 7 (Sustainable Development), SDG 11 (Sustainable Cities and Communities), SDG 13 (Climate Action) and SDG 17 (Partnerships for the Goals). The solution is based on a human-assisted AI model, which learns roof types at building level, local built and natural environment, geo-climatic datasets and community-validated intelligence to identify hyperlocal climate risk zones.

## Not Everyone Experiences Heat the Same Way – What Causes These Differences?

India is grappling with unprecedented heatwaves, with temperatures in the capital of Delhi exceeding 50°C. Over 75% of Delhi's total geographic area falls under its urban jurisdiction, with a population density exceeding 14,698 people per square kilometre, according to the 2011 census. Over 98% of the total population of Delhi (16.37 million people) reside in urban areas, as per the Delhi Heat Action Plan (Delhi Disaster Management Authority, 2023). Given this, there is a crucial need to establish the influence of urban morphology on heatwave intensity. Urban heatwaves exhibit spatial heterogeneity, and do not affect a city uniformly. There are certain conditions under which microclimatic zones within the city emerge as hotspots, experiencing significantly higher temperatures compared to surrounding areas. These disparities arise from a confluence of factors.

Firstly, densely built urban environments, characterised by high built-form density and extensive use of concrete and asphalt, exhibit high thermal retention compared to areas with open and green spaces. The use of such construction materials absorbs solar radiation throughout the day, releasing the stored heat slowly, and contributing to a persistent warming effect. Conversely, the vegetated areas with trees and grass provide crucial ecosystem services by promoting evapotranspiration, a natural cooling process (see Fig. 9.1). Secondly, the combined effects of shade from trees and evapotranspiration create cooler microclimates within urban landscapes. Consequently, the lack of green spaces and setbacks in densely populated areas exposes residents to a greater risk of heat stress, potentially impacting both physical and mental health.

ResSolv operates based on geospatial datasets, which comprise built characteristics, density and the environment of the study area. Based on roof-top architecture and spatial forms of physical infrastructure assets, the model monitors and assesses indicators of risk at the hyperlocal level within the urban context. Further, the validation of the simulated output is carried out through a ground-truthing assessment mechanism to achieve maximum accuracy in varied urban contexts.

There are specific atmospheric conditions that contribute to the prevalence of heatwaves. These include the presence of hot and dry air masses over a region, akin to a confined area of stagnant heat with no cooling mechanisms. Additionally, clear skies exacerbate the situation by eliminating the shade effect provided by clouds. Notably, the absence of moisture in the upper atmosphere further prevents heat dissipation. Prevalent wind patterns typically initiate heatwaves in North-west India, moving slowly towards the east and southern regions. These events can also develop independently under favourable conditions (hot, dry air, clear skies and trapped air) within any particular region (Indian Meteorological Department, n.d.).

Currently in India, the IMD provides resources for staying informed about heatwave threats at a national level. The department publishes daily bulletins with colour-coded heatwave warnings. Additionally, an open access interactive geographic information system (GIS)-based platform displays the forecast for the next five days. This gives visibility and builds awareness for areas or nearby regions that show a heatwave warning, which remains crucial in order to take steps to mitigate the impact. This also includes proactive preparedness measures for staying hydrated, avoiding strenuous activity during peak heat hours and prompting people to seek medical attention in case of discomfort. While national-level resources like the IMD's heatwave warnings are essential for raising awareness, pinpointing risks at a more granular level is crucial for taking targeted action. This is where hyperlocal AI solutions come in.

The ResSolv dashboard classifies risk hotspots at individual building and postal code level, with actionable outputs covering science- and evidence-based planning, which sharply target response, recovery and capacity building for city authorities, disaster management authorities, public infrastructure (roads, rail, power) agencies and corporate enterprises with critical business infrastructure. Plug-and-play services, such as integration with existing citizen interfaces, are under development.

## Heatwaves are Dangerous, Especially for Those Living in Crowded Areas

India has confronted a concerning new normal, with lethally high temperatures through April and May for three consecutive years. The temperatures recorded in the years 2022, 2023 and 2024 were abnormally high and in line with the heat projections (Coleman, 2024). Compared to the previous years, the variations of extreme temperatures recorded were spreading through a larger geographic extent, and peaks occurred earlier than anticipated. Alongside the early onset of heatwave occurrence, the duration is increasing concerns. There could be a potential period of overlap between the summer and monsoon seasons, which is particularly threatening, to both people and the environment. Prolonged exposure to extreme temperature coupled with humidity, especially for the people engaged in work in exposed settings and for special vulnerable groups, can be detrimental.

There are various factors (environmental and physical) that determine individual susceptibility to heatstroke and heat stress. The first factor is the weather. This includes how hot the surroundings are (temperature), moisture in the air (humidity), sunlight and wind conditions. The second factor is age, with underlying health conditions and fitness level influencing heat tolerance. In most cases, older adults, children and individuals with chronic health issues often have a reduced ability to regulate body temperature. This includes fitness level, as regular exercise enhances heat acclimatisation, improving heat tolerance for the human body, allowing it to sweat and cool down more effectively.

The physical environment where one resides can influence heat stress (see Fig. 9.1). The crowded areas essentially create smaller heat pockets, making them even more uncomfortable and dangerous during a heatwave. The people living in crowded areas face problems such as lack of ventilation, limited space, air pollution and less shade. Cities tend to be several degrees hotter than surrounding rural areas. This phenomenon, called the 'urban heat island effect', is caused by the factors mentioned above. This can be especially dangerous for the elderly, pregnant women and children, who are already more vulnerable to heat (UNICEF, n.d.).

The extreme weather conditions increase existing vulnerabilities faced by marginalised populations (World Weather Attribution, 2024). The internally displaced population, informal dwellers and migrant groups are disproportionately impacted by these extreme weather events. As is evident, the precarious living conditions amongst densely packed informal settlements, often lacking adequate ventilation and cooling infrastructure, create hotspots that trap heat and expose low-income communities to a greater risk of heat stress (UNDRR, 2023). This phenomenon underscores the social discrimination amplified by climate change, where those with limited resources bear the brunt of its consequences.

These neighbourhoods are characterised by limited access to essential resources and an environment that amplifies heat stress. Furthermore, the construction of homes using inexpensive, heat-trapping materials combined with the densely packed layout creates a microclimate that intensifies heat exposure at both the household and neighbourhood level. Additionally, limited purchasing power restricts access to air conditioners and coolers, common methods for alleviating heat discomfort. Finally, these communities often have limited access to affordable and effective health care, hindering their ability to cope with heat-related illnesses. As a result of these combined factors, low-income neighbourhoods, especially informal settlements, transform into heatwave hotspots, placing residents at greater risk of heatstroke and other severe health complications. Heatwaves act as an indirect factor that traps residents of informal settlements in a cycle of poverty, affecting their livelihoods. Sometimes, staying indoors seems like the only option, but indoor heat can be dangerous too.

AI-based modelling brings unprecedented levels of efficiency in forecasting and warning. Its focus on combating climate change for the urban low settlements, with early impact warnings to keep them and their neighbourhoods safe, as well as for the ecosystem of authorities and relief and response partners, with pre-emptive action to prevent avoidable losses and expedite timely relief during disasters, is a significant improvement over traditional systems operating these functions.

## Buildings Trap Heat!

Every heatwave occurrence in the world is now made hotter and more likely due to climate change. The months of the transition to the summer season have been abnormally hot in Delhi and other parts of India in recent years. For 26 days of April 2024, extreme heat conditions were recorded over larger geographies in the country, at least in smaller pockets within settlements, according to the IMD.

While air temperature plays a vital role, it offers an incomplete picture of thermal discomfort during heatwaves. The Heat Index addresses this by incorporating both temperature and humidity, providing a more accurate representation of how hot it feels to the human body. Ideally, our bodies maintain a core temperature between 37°C and 38°C. However, high humidity impedes sweat evaporation, the body's primary cooling mechanism. This can make individuals feel significantly hotter, even at seemingly moderate temperatures. For instance, on a humid day, 37°C can feel as scorching as 49°C, dramatically increasing the risk of heat stress (Delhi Disaster Management Authority, 2024).

Buildings, especially crowded ones, can trap heat, increasing indoor temperature and humidity levels. Hence, advisories recommending people to stay indoors should factor in the relative humidity of the particular region. With high indoor humidity, the 'feels like' temperature can be much higher than the actual reading, putting the occupant's health at risk. This is why it is crucial to consider the Heat Index, which combines temperature and humidity, to get a true picture of the heat's impact. The health impacts of heatwaves typically involve dehydration, heat cramps, heat exhaustion and/or heatstroke. Even when staying indoors during a heatwave, one must find ways to cool down inside, to avoid the dangers of high indoor heat and humidity. The scale and nature of the health impacts of

**Fig. 9.2.** Heat traps: indoor-outdoor temperature variation and approaches to tracking heat. Content authors' own. Illustration by Arundhati Hakhu.

heat depend on the timing, intensity and duration of a heat event, and the level of acclimatisation and adaptability of the local population, infrastructure and institutions to the prevailing climate (World Health Organization, 2024).

Buildings in cities absorb heat all day. This heat gets released slowly at night, keeping city nights warmer than usual (see Fig. 9.2). Because of this lack of night-time relief, the next day feels even hotter, and the peak temperature might hit earlier and last longer. This urban heat island effect (Ehsan *et al.*, 2021) extends the suffering of people from heat stress, particularly in vulnerable zones such as congested low-income neighbourhoods that cannot afford advanced technological interventions/strategies to combat harsh weather. This implies that buildings themselves can also make things worse. In addition, cramped spaces with poor ventilation and building materials trap heat, with poor insulation making it hard to regulate indoor temperatures. For example, from a physical sciences perspective, well-insulated and airtight dwellings retain heat from internal and solar gains, both in the winter and summer. Various spatial modelling and monitoring studies (*The Developer*, 2024) have shown that modern housing can overheat more than the vernacular style of architecture (i.e. traditional housing with natural cooling and ventilation spaces, built according to the local climatic conditions and available raw materials). Therefore, considering the indirect nature of existing epidemiological evidence, it is vital to define the basis of understanding of how indoor temperatures correlate with a health-based overheating threshold (Vardoulakis *et al.*, 2015).

Hence, determining the indoor Heat Index with indoor temperature and humidity measurements becomes so important. By installing data loggers and monitoring indoor temperature and humidity variations, one can identify the conditions that may lead to heat stress at an early stage. This is crucial in protecting people, particularly those in crowded buildings or poorly designed spaces. With this information, one can take steps to cool down indoor spaces before the heat becomes unbearable and puts people's health at risk.

## Which Sections of Your City are Heat-trapped?

Just like a fingerprint, every place is unique when it comes to how prepared it is for climate risks like heatwaves. There is ample data, analytics and insights based on geospatial datasets and secondary data sources to study floods, storms, heatwaves or earthquakes. However, physical structure (or built environment) data and evidence-based ground truth are missing elements in current climate risk and physical risk management needed to define the unique 'fingerprint'. Resilience AI uses artificial intelligence, machine learning and satellite images to create this fingerprint of an area's climate risk. It considers various roof typologies, building materials and spatial layouts to better understand heat build-up in different neighbourhoods. By analysing all these elements, the tool pinpoints areas at greater risk from heatwaves. This helps understand exactly which neighbourhoods and buildings are most likely to be affected, allowing us to take targeted actions to keep people safe during extreme heat. The next time a heatwave hits, these tools can help various local bodies to focus their resources on the communities that need it most.

Especially in poorly designed buildings, we must account for the building's overall energy efficiency to gauge heat risk at a hyperlocal/building level. This is where understanding a building's embodied and operational energy becomes vital. The embodied energy is used to create the building materials and construct the building itself. From observation, buildings with high embodied energy often use materials that trap heat, like concrete. The operational energy is used to run the building daily, including heating, cooling and lighting. There is an increased energy requirement for cooling buildings that are poorly insulated or ventilated, especially during heatwaves. By determining and mapping the embodied and operational energy of buildings, we can identify areas that can be improved by using materials with lower embodied energy and regulate indoor temperatures naturally through better insulation and ventilation and lower operational energy use. AI tools help by identifying buildings with high heat exposure and thus higher operational energy requirements. Some studies have explored the use of passive cooling strategies as pre-emptive measures before heatwave events. These strategies include the implementation of moisture-desorption cooling utilising hygroscopic materials to mitigate indoor heat stress in urban areas. The combined approach has demonstrated promising results, achieving a reduction of 0.43°C in indoor temperature through pre-cooling techniques, with an additional 1.31°C decrease by incorporating hygroscopic materials (Zhou *et al.*, 2020). To reduce indoor heat stress (Zhou *et al.*, 2020) and lower building energy consumption during heatwaves, passive and energy-efficient cooling techniques are required in the context of improving building energy efficiency. Thus, mapping a building's energy use is a crucial step towards creating heat-resilient cities.

ResSolv focuses primarily on climate adaptation in its current form. However, it has future potential for use in identifying local and hyperlocal mitigation actions with algorithm modifications. For example, the forecasting feature for indoor impact of extreme heat that currently identifies the most vulnerable families also has the potential to identify energy consumption surge for cooling in the same heatwaves, and to trigger actions to control energy demand spikes in the short term through financial incentivisation, and in the long term through focused passive cooling measures.

Effective heat mitigation necessitates a two-pronged approach: individual awareness and community preparedness (see Fig. 9.2). On an individual level, the installation of indoor data loggers can empower residents to track personal thermal conditions, allowing for optimised cooling strategies and improved health outcomes during heatwaves. Complementing this, at the neighbourhood level, automatic weather stations (AWS) provide hyperlocal climate data that can be used to trigger early heat warnings, inform urban planning for heat reduction and contribute to research efforts – ultimately fostering a more resilient community.

## What is Your City Doing to Deal with Heatwaves?

Heat action plans are a key strategy for enhancing the heat resilience of vulnerable populations in India, particularly during severe heatwaves. There is coordination among various subdivision and district levels and various stakeholders in preparation of heatwave guidelines and heat action plans (see Fig. 9.3). Rapid urbanisation has led to an increase in the population exposed to climate hazards. This, alongside mushrooming informal settlements, a rise in energy demands and shrinking green spaces, puts a strain on existing heatwave plans. While strategies like cool roofs, nature-based infrastructure and climate-risk informed building codes have been implemented across various cities, limited attention is given to retrofitting existing structures, particularly those with inadequate infrastructure, to enhance their liveability during heatwaves (World Weather Attribution, 2024). AI-based roof-top detection and hyperlocal risk assessment provides a heatwave risk score for individual housing

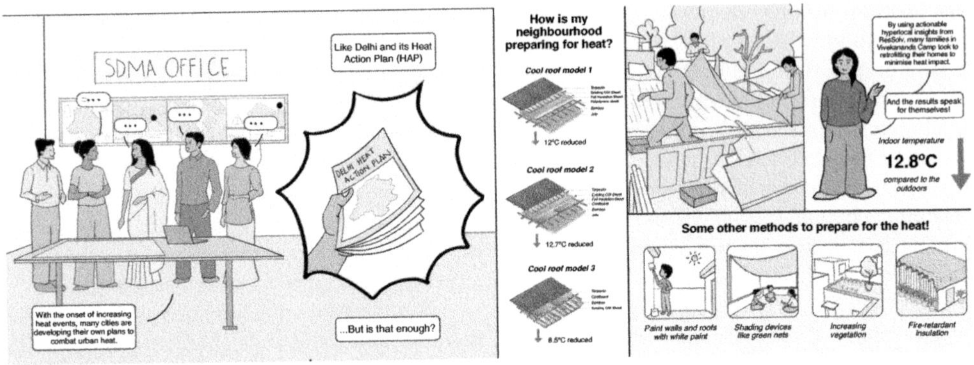


**Fig. 9.3.** City- and neighbourhood-level heatwave preparedness strategies. Content authors' own. Illustration by Arundhati Hakhu.

structures. Based on this, local communities, government organisations and aid agencies are informed of the associated threat, enabling them to take necessary actions. The platform seamlessly integrates with the state, district and city-level dashboards, thereby connecting emergency responders, public health officials, community leaders and residents. This enables inter-agency communication and coordination during heatwave events, thereby streamlining the development of heat action plans by providing a shared space for strategising and outlining roles and responsibilities. The heat action plans drafted for Indian cities must be further expanded to include mandatory regulations for the workplace. Especially for workers engaged in extended periods of outdoor activities, it is crucial that interventions like scheduled breaks and rest-shade-rehydrate (RSH) programmes be included as part of worker protection guidelines in affected regions (World Weather Attribution, 2024), especially in rural areas. This would substantially improve productivity and ensure protection from extreme heat conditions.

Heatwaves pose a serious threat to public health, particularly for vulnerable populations. To effectively combat these threats, cities need robust heatwave advisories and development plans. Using the risk assessment tool, city administrators will be able to pinpoint high-, medium- and low-risk areas within a city at the building level. This data-driven approach allows for targeted resource allocation, where heatwave hotspots over specific pockets of a neighbourhood have tailored interventions and advisories disseminated. AI-based tools like ResSolv can help identify an area as high risk, enabling authorities to prioritise deployment of cooling centres, medical supplies and outreach programmes to that specific location. This ensures resources are used most effectively to protect those most at risk. It simplifies the integration of heatwave preparedness measures into training modules for emergency responders and healthcare professionals. This ensures everyone has the knowledge and skills to handle heat emergencies effectively. Additionally, it can be used to disseminate real-time heat advisories and public awareness campaigns directly to residents. Residents can access this information easily, allowing them to take necessary precautions to stay safe during heatwaves.

Current heatwave plans in the country prioritise public health, but that might not be enough. As global warming worsens, heatwaves will become more severe and widespread. To effectively manage these intense heat events, local governments will need to use technology to assess risks at a very granular level. By leveraging new technologies, cities can move from reactive to proactive heatwave management, safeguarding the health and wellbeing of their citizens, particularly the most vulnerable populations.

## How is Your Neighbourhood Preparing for Heat?

There is a growing need for taking a proactive approach to heatwaves. Communities and local government bodies can identify buildings most susceptible to heat, allowing for targeted interventions. As an example, the high-risk structures were identified in Vivekanand Camp, New Delhi, an informal settlement that has successfully tackled heat with 'cool roof' interventions. Here, three models were prototyped and piloted using readily available and recycled materials. These cool roofs were incredibly effective, with some models bringing down indoor temperatures by 12°C (see Fig. 9.3). This approach offers several long-term benefits. Firstly, it prioritises the wellbeing of residents, particularly those at high risk from heatwaves. Secondly, by promoting localisation by utilising materials available and recycled on-site, the solutions are effective and account for environmental sustainability. Finally, fostering community collaboration strengthens social cohesion and empowers residents to take ownership of their environment.

Building-level heat risk assessment would provide valuable insights that enable scalable and impactful heatwave preparedness strategies across diverse communities. As more neighbourhoods embrace this data-driven, collaborative strategy, the collective impact on mitigating heat risks and ensuring resident wellbeing will be significant. Through scalable solutions like cool roof designs and the documentation and periodic analysis of their thermal performance data, other communities can be inspired to replicate these strategies. Further, through the disseminated information and coordinated volunteer efforts, residents can be empowered to take ownership of mitigation strategies within their neighbourhoods. This demonstrates the potential for neighbourhoods to implement similar cost-effective solutions like cool roofs, white paint, shade netting and increased vegetation.

## How are You Preparing for a Heatwave?

While there is increasing recognition of the need for tools for climate action, adaptation and resilience, the use of these largely remains clustered at the city scale. Localised climate security for local communities and organisations receives limited attention. Effective planning for climate hazards can be pivotal in reducing the cost of this inaction, as well as the other social, physical and psychological costs associated with losses in the event of a disaster. Heatwaves pose a significant threat, and effective planning can significantly reduce their impact. Individual actions are a critical practice in the wake of such rising climate-induced hazards. It empowers users to assess the readiness of their surrounding built environments to withstand climate-induced disasters, take aligning action and become agents of change in their local area.

Using the heatwave risk alert, a user can assess the underlying geo-climatic risks that threaten their lives and livelihoods and safeguard their moveable and immovable assets to minimise losses. Through climate innovation and technology, a user can obtain precise and customised risk assessments for any area at the micro level. Additionally, using this data, an individual can gain hyperlocal and action-oriented insights into the climate risks that they face in their built environment. This knowledge can greatly contribute to building community-level preparedness, especially in high climate-risk areas, and reduce losses very significantly. The process of community-based disaster preparedness (Assam State Disaster Management Authority, 2013) helps in mobilising vulnerable populations in an organised way to achieve a safe and resilient community. The end goal is a dynamic community that is capable of undertaking individual and group tasks to address and recover from hazard events, that unites the group cohesively during decision making, handles conflicts and resolves issues. AI-based tools like ResSolv have the potential to play a vital role in this individual preparedness (see Fig. 9.4). Dissemination of real-time heat advisories, allowing residents to tailor their actions based on the severity of the event, can make a very significant difference.

## Using AI for Solving the Challenge of Extreme Heat Risk

AI-based climate risk assessment tools can help by blending built environment experience and community behaviour with cutting-edge data analysis capabilities to deliver climate resilience at scale. The extreme heat trends observed across the country and globally provide yet another reason to invest in and promote such technological

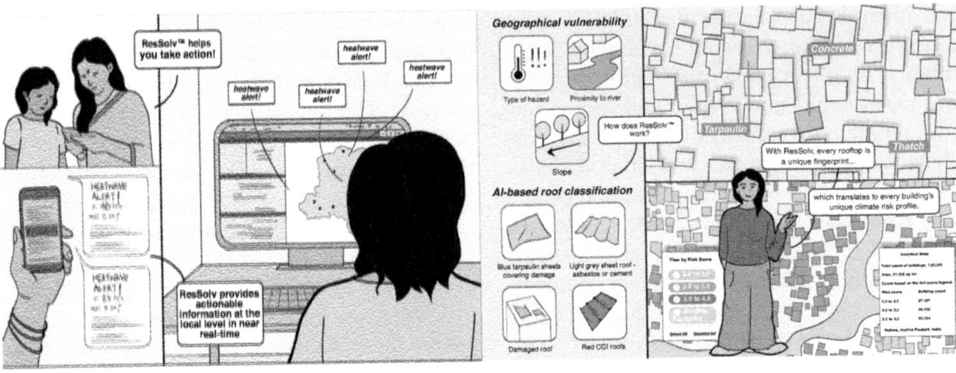


**Fig. 9.4.** Know your risk with ResSolv™: Proactively mitigate heatwaves. Content authors' own. Illustration by Arundhati Hakhu.

solutions. It is essential to explore the multifaceted challenge of heatwaves and the potential solutions that lie at the intersection of cutting-edge technology, community engagement and individual action. A key takeaway is the importance of hyperlocal, data-driven heat risk assessments and customised actionable advisories to address extreme heat conditions at scale.

AI offers tools with the ability to analyse geo-climatic data and built-environment characteristics, empowering communities to identify high-risk areas (see Fig. 9.4). This information can then be used to guide frugal interventions such as the use of cool roof materials or a community health response, maximising impact while minimising cost. The flexibility of the application creates significant potential for scaling in various geographical typologies. The approach is versatile and is based on commonly available hardware without any high-end specifications. Within India, the tool has been piloted in 17 cities covering 100,000 buildings across 500 km$^2$, covering customers on the eastern coast experiencing increased cyclones, on northern terrain experiencing heat, in hilly regions exposed to earthquake risk and on the southern coast experiencing floods. The implementation partners and stakeholders include State Disaster Management Authorities, grassroots organisations and ecosystem partners.

Impact warning systems can play a crucial role in mitigating heatwave risks. By leveraging insights and integrating them with existing infrastructure, communities can receive timely alerts and actionable advisories, allowing residents to take necessary precautions. However, effective heatwave preparedness requires a human-in-the-loop approach. The usage of AI should be viewed as a tool to augment human action. Building heat resilience necessitates action at various scales, from regional to city to local at individual level. By acting at all levels, we can create a safer and resilient future for all in the face of a changing climate.

## References

Assam State Disaster Management Authority (2013) *Community-Based Disaster Preparedness Course Book*. Doctors For You. Available at: https://asdma.gov.in/download/modules/Community%20Based%20Disaster%20Preparedness-Peripheral%20Level.pdf (accessed 16 July 2025).

Centre for Science and Environment (2023) Climate India 2023: An assessment of extreme weather events. Available at: https://www.cseindia.org/india-2023-extreme-weather-events-11973 (accessed 12 July 2024).

Coleman, J. (2024) Chance of heatwaves in India rising with climate change. Nature. Available at: https://www.nature.com/articles/d41586-024-01577-5 (accessed 16 July 2025).

Delhi Disaster Management Authority (2023) Delhi Heat Action Plan. Available at: https://ddma.delhi.gov.in/sites/default/files/ddma/generic_multiple_files/nidm-heat_wave_action_plan_05.07.2023_1.pdf

Delhi Disaster Management Authority (2024) *Heat Action Plan 2024-25*. Delhi Disaster Management Authority, Delhi.

Ehsan, S., Abbas, F., Ibrahim, M., Ahmad, B. and Farooque, A.A. (2021) Thermal discomfort levels, building design concepts, and some heat mitigation strategies in low-income communities of a south Asian city. *International Journal of Environmental Research and Public Health* 18(5), 2535. DOI: 10.3390/ijerph18052535.

Indian Meteorological Department (n.d.) FAQ on heat wave. Available at: https://internal.imd.gov.in/section/nhac/dynamic/FAQ_heat_wave.pdf (accessed 16 July 2025).

National Ocean and Atmospheric Administration (n.d.) Wet Bulb Globe Temperature. National Weather Service. Available at: https://www.weather.gov/ict/WBGT (accessed 16 July 2025).

Ravindra, K., Bhardwaj, S., Ram, C., Goyal, A., Singh, V. *et al*. (2024) Temperature projections and heatwave attribution scenarios over India: A narrative review. *Heliyon* 10(4), e26431.

The Developer (2024) Are energy efficient buildings more likely to overheat? Not if you do it right. The Developer. Available at: https://www.thedeveloper.live/opinion/opinion/are-energy-efficient-buildings-more-likely-to-overheat-not-if-you-do-it-right (accessed 16 July 2025).

UNDRR (2023) 'It's All Our Burden': Poorest Women Hardest Hit by Heatwaves in India. PreventionWeb. Available at: https://www.preventionweb.net/news/its-all-our-burden-poorest-women-hardest-hit-heatwaves-india#:~:text=Heat%20waves%20put%20an%20inordinate,ways%20to%20cool%20their%20families (accessed 16 July 2025).

UNICEF (n.d.) Heat Wave Safety Tips. Available at: https://www.unicef.org/parenting/emergencies/heat-wave-safety-tips (accessed 16 July 2025).

Vardoulakis, S., Dimitroulopoulou, C., Thornes, J., Lai, K.-M., Taylor, J. *et al*. (2015) Impact of climate change on the domestic indoor environment and associated health risks in the UK. *Environment International* 85, 299–313. DOI: 10.1016/j.envint.2015.09.010.

Word Bank (2021) *Climate Risk Country Profile: India*. World Bank, Washington, DC. Available at: https://climateknowledgeportal.worldbank.org/sites/default/files/country-profiles/15503-WB_India%20Country%20Profile-WEB.pdf (accessed 16 July 2025).

World Health Organization (2020) Heawaves. Available at: https://www.who.int/health-topics/heatwaves#tab=tab_1 (accessed 12 July 2024).

World Health Organization (2024) Heat and Health. Available at: https://www.who.int/news-room/fact-sheets/detail/climate-change-heat-and-health (accessed 12 July 2024).

World Weather Attribution (2024) World Weather Attribution. Available at: https://www.worldweatherattribution.org/climate-change-made-the-deadly-heatwaves-that-hit-millions-of-highly-vulnerable-people-across-asia-more-frequent-and-extreme/ (accessed 16 July 2025).

Zhou, X., Carmeliet, J., Sulzer, M. and Derome, D. (2020) Energy-efficient mitigation measures for improving indoor thermal comfort during heat waves. *Applied Energy* 278, 115620. DOI: 10.1016/j.apenergy.2020.115620.

# 10 Tokyo's Perpetual Resilience Project: Between Local Knowledges and Universal Modernist Concepts

**Christian Dimmer***
*Waseda University, School of International Liberal Studies, Tokyo, Japan*

**Abstract**

This text explores Tokyo's evolving urban resilience strategies in response to disaster risks and climate change, with a particular focus on the recently launched Tokyo Resilience Project (TRP). Released to mark the centennial of the 1923 Great Kantō Earthquake, TRP outlines five key threats while presenting a long-term vision for a safe, sustainable Tokyo over the next century. The chapter situates Tokyo within its historical context, arguing that a distinctive 'disaster culture' developed in the pre-modern era, in which disasters were accepted as inevitable. Thus, responses focused more on rapid rebuilding than on preventive planning, fostering localised resilience and community agency. In contrast, the modern period introduced a shift towards state-led, engineering-based approaches. While technically advanced, these have inadvertently undermined grassroots resilience by displacing citizen engagement with expert-driven, technocratic solutions. The TRP is critically examined for its limited civic participation, its top-down orientation and its lack of cross-departmental coordination. Such fragmented, depoliticised approaches risk fostering single-solution responses that are ill-equipped to address complex, 'wicked' problems. Drawing on ecological systems theory, the chapter advocates a paradigm shift towards adaptive, participatory governance models, which embrace uncertainty and empower local communities. Influenced by Jane Jacobs's ecological idea of urbanism, the chapter highlights interdependence, redundancy and bottom-up innovation as key features of resilient urban systems. Ultimately, resilience is not only a technical issue but also a political and epistemological one. The chapter concludes by calling for the revitalisation of civic agency and community-based resilience as essential tools in navigating an increasingly uncertain, climate-challenged future.

## Introduction

On 1 September 1923, the Great Kantō Earthquake devastated vast parts of Tokyo and Yokohama. Around 140,000 people were dead or missing and many hundreds of thousands more lost their homes and livelihoods. There are few cities in the world that are exposed to a similarly high disaster risk than Tokyo, with experts predicting with a 70% probability that the city will be struck by another major earthquake within the next 30 years. The Tokyo Metropolitan Government (TMG) took the 100th anniversary of the Great Kantō Earthquake as an opportunity to launch its new 'Tokyo Resilience Project', which is seeking to strengthen the city against five identified disaster threats and aims at providing its citizens 'safety for the next 100 years' (TMG (Tokyo Metropolitan Government), 2023a).

The release of the project coincided with a flurry of other bold-sounding policy papers like

*Corresponding author: christian.dimmer@waseda.jp

 *Climate Change and Risk Mitigation: Reducing Vulnerabilities and Enhancing Resilience* (ed. P. Barnes)
DOI: 10.1079/9781800624153.0010

'Tokyo Sustainability Action' (TMG, 2023b), aimed at aligning the city with sustainability principles; the 'Tokyo Bay esG Project', promoting the utilisation of Tokyo Bay as laboratory for innovations allowing Japan to tackle pressing challenges like climate change; or the vision of a city for the next generation, 'SusHi Tech Tokyo' (TMG, 2023c), short for 'Sustainable High City Tech Tokyo', positioning Tokyo as model and leader for the global smart city market. All of these policy statements commonly use familiar 'global' buzzwords and seek to firmly root Tokyo as a positive force on the international stage in the fight against the climate crisis and other global challenges.

The objective of this chapter is to outline and evaluate strategic aspects of Tokyo's new resilience project. In order to achieve this, the chapter first provides an in-depth historical contextualisation of the city's development that took place under the constant threat of natural disasters and devastation, showing how a distinct 'disaster culture' emerged that permeated and informed policy making, planning and cultural practices in fundamental ways. Secondly, the newly released 'Tokyo Resilience Project' (TRP) is examined. Although the policy was inspired by the Intergovernmental Panel on Climate Change (IPCC) Sixth Assessment Report and is manifest of a growing urgency regarding climate change, it represents a response to wider environmental risks. Disaster precautions and responses to the following five threats are expounded: (i) floods and storms, (ii) earthquakes, (iii) volcanic eruptions, (iv) disruptions of power and communication and (v) infectious diseases.

A brief semiotic reading of the document and supporting texts raises questions about the underlying assumptions and biases in regard to resilience against the background of historical, pre-modern and early modern conceptions of its use. Thirdly, the chapter concludes with a theoretical reflection on ecological and engineering resilience, as well as on concepts like emergence and contingency for a critical assessment of Tokyo Met's resilience strategy.

I argue that in pre-modern Japan, disaster was profoundly accepted as a force outside of the full grasp of humankind and thus a high awareness of and resilience against calamities emerged. In modern Japan, as in any advanced nation that embraces modernist epistemologies and ontologies, and as exemplified by the TRP, man-made and natural catastrophes have come to be portrayed as manageable by experts and government authorities, leading to a disengagement of non-expert citizens. A return to a pre-modern notion that accepted the omnipresence and inevitability of disasters, combined with ecological ideas of resilience, would enable the urban system of Tokyo to transition into a more sustainable system state, better suited to perpetually adapting to a changing world.

While the dominating engineering-based ideas of resilience, as represented in the TRP, are perhaps adequate to enable specific urban functions to rebound and to work similarly as before a disturbance, they stand in the way of activating the collective problem-solving skills of an engaged, cooperative citizenry. These engineering-based notions keep the citizens disengaged from active adaptation thinking and acute risk awareness, as policy orthodoxies largely emphasise the ability and authority of experts to take care of preparing the Japanese city for coming disasters and other yet-unknown future threats.

## Edo/Tokyo's Pre-modern Tradition of Disaster Resilience

Few other large cities in the world were destroyed and rebuilt as frequently and on as sweeping a scale as Tokyo. Edo, as Tokyo was called until the opening of the country to the West during the 1860s, had always been exposed to recurring, large-scale disasters since the city's foundation in the early 17th century. Calamities were such an integral part of everyday life and thinking that a distinct 'disaster culture' emerged, with earthquakes, volcano eruptions, storms, floods and fires figuring prominently in cultural representations and practices.

After the Great Meireki Fire (1657), which claimed around one hundred thousand lives, the frequent conflagrations, regularly devastating parts of the largely wooden city, came to be referred to as 'flowers of Edo' (Kelly, 1994, p. 311; Tate, 2009). The emergence of this prominent trope is an indication that the devastation of the city had come to be normalised as a recurring part of urban life and that Edoites had no choice but to accommodate such events as outside of human control. Idioms in the Japanese language like *shikata ga nai*, or 'it cannot be helped', still echo this acceptance of the inevitable.

Art historian Gennifer Weisenfeld points out that 'disaster is a defining feature of Japan's cultural landscape, and, consequently, the country's general belief system has integrated the cyclicality of destruction and renewal' (Weisenfeld, 2012, pp. 13–14). It is fair to hypothesise, though, that this pre-modern *deep acceptance* of the seemingly inevitable might have led to a reluctance of planners and authorities of the past to fundamentally 'build back better', to use a contemporary slogan that gained prominence around the Third UN World Conference on Disaster Risk Reduction, which took place in Sendai in March 2015. The resulting Sendai Framework for Disaster Risk Reduction 2015–2030 emphasised the importance of building back better communities after disasters, by avoiding mistakes of the past and *simultaneously* focusing on resilience and sustainability in reconstruction efforts to adapt cities to climate change-induced long-term challenges.

Aside from the cultural dimension, the need to build tremor-resistant structures induced the development of post-and-beam wooden architecture that was relatively light and resilient to quakes but vulnerable to fire. Instead of an illusive modernist idea of preventing disaster altogether, focus in pre-modern Japan was on resilient, swift response to what was seen as outside of human control. Historian William Kelly states accordingly that firefighting was less directed at 'extinguishing fires ... but toward limiting the fire's further spread' (Kelly, 1994, p. 313). The disaster-resilient attributes of the city was thus a delicate trade-off: to lessen earthquake damage, wood and paper came to serve as main building materials, which in turn increased the vulnerability to fires. Thus, instead of seeking to avoid destruction, historically, an emphasis emerged that favoured quick post-disaster rebuilding, resulting in a *normalisation* of devastation in the collective psyche. Historian Edward Seidensticker stresses that the pride of Edo's people was a swift recovery, or in other words, their resilience. He states that 'the proper merchants insisted upon speed ... If a shop had not resumed business within three days, common mercantile wisdom held, it had no future' (Seidensticker, 1991, p. 8).

Often disasters don't occur as singular events but in compound, cascading form. Fire was the cause of death for the vast majority of the 120,000 victims of the Great Kantō Earthquake. These erupted after the tremors struck at 11:58 AM on 1 September 1923 — a time when midday meals were being prepared on coal-fired stoves in hundreds of thousands of homes — in a city where still only a fraction of buildings possessed degrees of fire-resistance and large green spaces that offered a safe escape were rare.

It is important to point out that beyond essentialist, cultural traits, other concrete reasons led to such thinking. Urban historian Carola Hein cites, for example, strong economic incentives for providers of cheap rental housing to cut costs for building materials, assuming the regular loss of their investment. To put it bluntly: Cheap building materials equaled higher profits and minimized losses when the next, inevitable disaster struck (Hein, 2005, p. 215). On the side of the powerful, there was a lack of capacity and a political reluctance to enforce stricter building codes and planning decisions that could have led to greater resilience against disasters, and thus recovery was left in the hands of private builders. Instead of any notion of 'building back better', in essence the pre-modern motto was swiftly 'building back as before', thus tacitly accepting the regular loss of life and property.

In other words, the acceptance of the inevitability of disasters in pre-modern Japan led to the emergence of a strong psychological, individual and collective resilience with socio-cultural protocols emerging that would allow for a quick re-establishment of the city *without* seeking to fundamentally alter it. Historian James McClain stresses that, although the pre-modern government of the Shogunate that theoretically held the supreme power in the city designated numerous firebreaks like wide boulevards or open spaces around bridges after major disasters, it was not able to keep these sites permanently vacant, as surrounding residents soon encroached on the land (McClain, 1994, p. 107). Thus, the city soon swung back to similar states of vulnerability. We will return to this point later, when discussing that the environmental challenges cities are facing today require massive, perpetual sustainability transitions for perpetuity along with the governance structures and mindsets that are needed to promote this.

## Modern Notions Overlay Pre-modern Ideas of Resilience

When a geo-strategically isolated Japan had to open its ports to the modern West in the 1860s and the old feudal socio-political order was supplanted by a modern nation state, modern influences and foreign knowledges and mindsets entered into the country.

Historian Gregory Clancey demonstrates how earthquakes in Japan shaped the country's 'techno-social' development (Clancey, 2006). He details that after an initial unchallenged adaptation of Western knowledge, the powerful Nobi Earthquake of 1891 shattered many modern, Western-style buildings and infrastructure in Gifu prefecture. As a response, Japanese engineers 'took on the task of adapting (the newly introduced) foreign knowledge to Japanese nature (and geography) and marrying the knowledge of the (modern) present to the knowledge discovered by their countrymen long ago' (Clancey, 2006, p. 2).

However, Clancey argues that in the aftermath of the Great Kantō Earthquake, the country's state of being an 'earthquake nation' was actively sublimated by elites in Japan who desired to disassociate themselves from the weakness that was perpetually posed by disasters. On the other hand, foreign visitors were less interested to learn from disaster-resilient building structures and cities. Instead, they were captivated by finding a supposedly ahistorical, essentialist Japanese architectural modernism that confirmed their ideologies of universally applicable modernist principles. 'The characterisation of Japan as an earthquake nation, which had helped construct faith in Japanese science internationally in the Meiji period, became ... a potential source of Japanese embarrassment in that same arena. The sudden, all-too-complete destruction of their capital threatened to erase overnight the image of the Japanese as a people *in control* of their own nature' (Clancey, 2006, p. 226, my italics).

Without specifically referring to Japan, anthropologist James C. Scott points out that converse to the pre-modern world, under high modernism 'the scope of [state] intervention was potentially endless' (Scott, 1998, p. 92). While pre-modern societies remained rather stable, as seen above with reference to the failure to preserve firebreaks in Edo, under high-modernism 'society became an object that the state might manage and transform with a view toward perfecting it. A progressive nation-state would set about engineering its society according to the most advanced technical standards of the new moral sciences' (Scott, 1998, p. 92). Thus, in respect to earthquake proofing, society in modern Japan came to trust a technocentric, professionalised approach to safety by engineered design. It is fair to assume that climate change adaptability will be seen in a similar light.

Accordingly, Clancey states that 'Japanese seismology and earthquake engineering would continue to grow in skill, application, and prestige in the postwar period (after 1945) but mitigating the impacts of earthquakes and other natural hazards became a technical concern with less obvious cultural meaning or import' (Clancey, 2006, p. 233).

In other words, while resilience to disasters had *deep* pre-modern roots and has *broadly* influenced culture and society, it became a domain of social engineering, state management and a reliance on expert knowledge that took over the task of perpetually transforming urban Japan under the pretext of disaster-proofing (Dimmer and Kammerbauer, 2024, p. 98). This reduced society's resilience overall as disaster awareness and preparedness were no longer broadly *and* deeply rooted. Confidence in the individual ability to 'bounce back' and collective capacity to rebuild swiftly waned, as disaster preparedness moved from individual lived experience into the domain of the modern state with its expert engineers and its restless bureaucracy. The results of this can be studied in the areas hit by the 2011 Great East Japan Earthquake as well as the recent earthquake on the Noto Peninsula in January 2024. Here, self-help was far less intensive than in early modern times, and citizens have been passively waiting for the government to act, with the result of a delayed, disjointed reconstruction process (Dimmer, 2018).

To go further, it can be argued that the unfettered trust in the work of the engineers and science created new, unprecedented vulnerabilities. Environmental historian Philip C. Brown makes this point with reference to the extensive damage from the tsunami disaster that resulted from the 2011 Great East Japan Earthquake: 'People's faith in dikes and other (disaster protection) projects' ability to create a safe environment' (Brown, 2014, p. 10) led to blind trust in experts and engineers, who in turn had faith that technology could master any challenge nature would pose. Based on the study of historical and contemporary disasters, Brown stresses that the 'long-term unintended consequences of programs designed to solve one set of (narrowly defined) issues without awareness of links between the problem area and other (interdependent) systems' increased the overall vulnerability of cities (Brown, 2014, p. 2).

The great tsunami breakwater in the North-east Japanese town of Kamaishi exemplifies this. The massive engineering structure not only failed to protect the city itself but also redirected the destructive tsunami powers to less protected neighbouring towns, which were completely destroyed (Onishi, 2011). The Fukushima nuclear disaster of 2011 was often referred to as *sōteigai*, or 'beyond expectation' (Krolicki *et al.*, 2011), indicating an act of god, outside of human control. What science and technology studies scholars call a 'black swan event' revealed how engineers had been blindsided by their belief in the infallibility of their mathematical models. Later, it also emerged that the plant operator TEPCO's own safety engineers had warned that similar tsunamis had struck the region regularly in previous centuries, but incorporating such inconvenient knowledge into risk mitigation efforts would have caused huge costs for TEPCO. A supposedly unanticipated act of god turned out to be a mundane act of cost cutting, coupled with a reluctance to appropriately anticipate future hazards.

In summary, with the transition from premodern Japan to modern Japan, the faith in the ability of engineering to prevent[1] the impacts of disasters has gradually increased and, at the same time, the ability of citizens of communities to actively anticipate risk and rebuild quickly through their own initiative has begun to wane. The more Japan turned into an advanced, modern, bureaucratic state, the more rules and regulations came to isolate society from active engagement with disasters.

## Gotō Shimpei's 1923 Reconstruction Plans as a Template for Today?

On 1 September 1923, the Great Kantō Earthquake struck Tokyo and Yokohama, with roughly 140,000 dead or missing. In Tokyo, about 45% of the urban area was devastated by the tremor as well as the large fires that followed in its wake. The picture of the destruction reflected important spatial characteristics of the city: the fires were most destructive and deadly in the densely built-up East Tokyo, the so-called 'low city' or *shitamachi*, where the commoners had lived during the feudal period. Conversely, the greener, less densely populated former residential districts of the samurai class in the hilly Western part — the so-called 'high city' or *Yama-no-te* area — escaped mostly unscathed. One lesson for urban planners was that large urban green spaces like Ueno Park happened to serve as barriers that stopped the spread of the flames and later, as temporary shelters for hundreds of thousands who had lost their homes.

Historian Charles Schencking examined how the former colonial official Gotō Shinpei took charge of the reconstruction effort and swiftly proposed a grandiose scheme that came to be mocked as the '800 Million Yen Plan'. Gotō sought to remedy the challenges of a premodern city that had been planned around foot traffic and that was hardly able to fulfil the needs of a modern industrial society. As urban geographer André Sorensen stresses, it is a recurring motif in *modern* Japan's planning history that ambitious plans for a 'building back better' with massive green belts, parks, boulevards and brand new infrastructure were gradually pared down due to resistance from the political elites with competing visions and from a populace hoping to quickly return to their personal predisaster normalcy (Sorensen, 2002).

Similar to disaster reconstruction efforts in the past and in other parts of the world, Gotō and those in favour of far-reaching urban transformation saw the earthquake as an 'ideal opportunity to make a perfect new city' (Gotō in Schencking, 2013, p.154), to build back a better, truly modern Tokyo. However, it turned out that planners underestimated the complex socio-political realities, and mostly only those projects that directly served economic development were realised. While André Sorensen considered the reconstruction project as mostly positive, allowing for the first time implementation of modern planning techniques on a large scale (Sorensen, 2002, p. 350), Schencking highlights the severe struggles between the ruling elites and the failure to achieve a far more thorough urban reform (Schencking, 2013). Today, Tokyo faces a similar challenge to rapidly adapt to climate change. Similarly ambitious projects have been proposed by the planners and engineers; however, the fragmented governance landscape that derailed the 1923 reconstruction project is even more complex today, yet lessons from the governance failure have not been sufficiently drawn.

Although reconstruction plans for Tokyo were severely curtailed, with the initial budget for public parks for example being reduced by 90%, the rebuilding project was officially completed after only 6 years in 1930 and successfully modernised at least the central-east part of Tokyo. For the first time, it established a rational street system with footpaths, modern bridges, small squares, fire-proof schools, public housing and some prominent public parks, in places where little open space had previously existed (Sorensen, 2002, pp. 125–133), although already in 1930 many observers, among them the emperor himself, bemoaned that a more far-reaching transformation could not be achieved (Schencking, 2013, p. 305). And indeed, the reconstructed parts of the city turned out only insignificantly less vulnerable, when the American air force targeted Tokyo in March 1945 with its firebombing campaign (Schencking, 2013, p. 305–307).

However we choose to evaluate the success or failure of the Tokyo reconstruction project, to transform the existing Japanese city under the objective of preparing it for the next, inevitable disaster, emerged thereafter as one of the central driving forces for urban planning in Japan. To return to the above quote by James C. Scott, 'the scope of intervention was potentially endless' (Scott, 1998, p. 92) and reconfiguring the existing Japanese city against disaster became an unchallenged, perpetual driver for urban change, relentlessly pursued across all state agencies and institutions. With every new major disaster, stricter precautions in national legislation were introduced to minimise damage and casualties (Dimmer and Kammerbauer, 2024, p. 110), while at the same time, the unplanned, fragmented character of many parts of Tokyo frustrated a swift implementation of these planning goals. Furthermore, the need for perpetual renewal facilitated to the growth of powerful real estate corporations and strong vested interests that pushed for more and more redevelopment.

## The Tokyo Resilience Project

The recently issued TRP (TMG, 2023a) was consciously published at the occasion of the centenary of the Great Kantō Earthquake and proudly cites the visionary effort of Gōtō Shimpei, who sought to transform Tokyo from a still largely pre-modern urban structure built around foot traffic into a modern, industrial metropolis with cutting-edge infrastructure.

While only mentioned in passing in the policy document itself, the project website[2] prominently features the visionary plans of Gotō Shimpei in the aftermath of the 1923 Great Kantō Earthquake as a model for restructuring the city 'for the next 100 years'. Not mentioned are the many problems and controversies that led to what prominent contemporaries already described as the failure of Gotō's overly ambitious and ideal plans (Schencking, 2013).

A critical study of these reasons reveals that governance in Tokyo today is still plagued with many similar political, administrative and institutional impediments that frustrated the execution of Gotō's comprehensive urban reform ideas over a hundred years ago: these range from competing reconstruction priorities and disarray of different public authorities to a prioritisation of economic promotion over quality-of-life improvement, to name a few. Alas, the TRP website presents the successes of the 1923 reconstruction projects, like the construction of 52 ferro-concrete schools, serving as evacuation sites in future disasters, three large parks, etc., without reference to the many shortcomings of the project or proactively developing governance capacities to better prepare for future reconstruction.

In the colourful policy document, the Tokyo Metropolitan Government identifies the five major threats – (i) floods and storms, (ii) earthquakes, (iii) volcanic eruptions, (iv) disruptions of power and communication and (v) infectious diseases) – as it seeks to strengthen the city 'for the next 100 years'. The plan vaguely sets out a 'roadmap to achieving our vision for Tokyo in the 2040s' (TMG (Tokyo Metropolitan Government), 2023a). In the introduction, governor Yuriko Koike states that '2023 marks the centennial of the Great Kantō Earthquake. We will strongly promote this project, which will begin in this milestone year, to ensure that Tokyo, the capital of Japan, becomes a sustainable city that offers its residents peace of mind for the next 100 years to come' (TMG (Tokyo Metropolitan Government), 2023a). It is worth stressing that the pre-modern 'disaster culture' of Japan developed precisely because there was no peace of mind, but can be seen as a *memento mori* or coping mechanism against a looming threat. The resilience that the pre-modern Edoites and their communities displayed is a result of alertness that may be seen as the opposite to peace of mind.

Also, climate change and the IPCC Sixth Assessment Report of 2022 are stated in the English summary of the plan as pressing challenges the TMG plan is to address, but in the Japanese version there is no mention of the IPCC report. It is also worth pointing out that there have been five earlier IPCC reports that all painted a similarly bleak image of climate change-induced challenges for cities, and yet TMG had not responded to these earlier warning calls with an expected degree of urgency. This may suggest that it is important to TMG administrators to demonstrate to the international community that Japan's capital is readying itself for massive challenges while not finding it important to communicate this to domestic audiences. During the 'Urban resilience week', organised by the UN in October 2023, Tetsuro Fujisaki, then-Director for the Tokyo Resilience Project at TMG, touted the new policy, for example, as a global model, addressing globally recognised issues (Fujisaki, 2023). It is also worth mentioning that only the unprecedented heatwave in summer 2023 triggered the first major public debate about climate change that even threatened to alter Japan's culture in profound ways (Alt, 2023).

Following the list of five disaster sources cited above, the TRP presents future visions and objectives for every scenario (TMG (Tokyo Metropolitan Government), 2023a), followed by a long list of concrete key initiatives and projects (pp. 47–147) with which TMG hopes to reach these. The final chapters deal with project promotion and implementation (pp. 181–183) and list the investments for the next 10 years and until 'the 2040s', but it is not clear what the precise target year is (pp. 185–186). A casual look at the structure of the document highlights the strong emphasis on specific projects, with little space allocated to discussing governance capacities or government coordination between municipalities within Tokyo Metropolis and between Tokyo and neighbouring cities and prefectures. Most of the featured projects are infrastructure projects. Also, funding, project phasing or theoretical underpinnings for new compound or cascading patterns of disaster under a scenario of accelerating climate change are hardly discussed beyond very basic statements.

The final pages of the document show 'the opinions of Tokyo citizens' (pp. 189–192), with TMG making the implicit claim that the preceding policies address the stated concerns of the citizens. It is noteworthy that the citizens explicitly appear for the first time at the very end of the document and that there seems to have been little more citizen engagement in the making of the plans than to collect 'opinions' through two week-long web surveys for adults and children between the end of October and November 2023. As a reminder, the TRP was published in December 2023, strongly suggesting that there was no meaningful citizen engagement. Lastly, the Japanese term *Tomin* is used to indicate the subject of the survey, broadly meaning 'residents of Metro Tokyo'. There is no further distinction provided in the survey, and indeed, the whole TRP does not offer specific considerations for the millions of commuters who are working and studying in central Tokyo, the rapidly growing share of elderly or the blossoming number of foreign residents, which quadrupled between 1985 and 2019 (Nippon.com, 2019).

Don't these demographics have specific needs and shouldn't these be accommodated in the plans? As with many planning policies in urban Japan, plans are made by experts on behalf of a politically passive citizenry and mostly present technical solutions to what Rittel and Webber have famously called 'tame problems' (1973). Demographic and governance considerations are under-represented.

A motif of the document are statements like, 'residents can live their lives without feeling worried about flood and storm damage' (TMG (Tokyo Metropolitan Government), 2023a). We sense here a paternalistic attitude by a government providing safety to the citizens, who are to live their everyday lives without any worry, concern or, yes, constant awareness of potential dangers. This is precisely the attitude that Brown (2014) has criticised, as discussed above, as it discourages the development of a critical awareness of growing hazards and frustrates risk mitigation and disaster preparedness efforts on an individual and community level. It also undermines local autonomy and meaningful participation.

The long list of resilience projects between pages 42 and 159 of the plan seems to impress on the reader that the TRP doesn't exhaust itself in empty rhetoric, but that it offers specific and proactive measures. On closer examination, though, it appears that a number of the listed projects are not contemporaneous with the new policy document. Instead, many projects were developed sometimes decades earlier for their own reasons and by different government authorities and are now presented as part of a supposedly comprehensive, well-adjusted transition strategy. For example, the fire-proofing of neighbourhoods with few open spaces and a concentration of old, wooden buildings has been a priority task for over 50 years, and at the same time, its implementation has always been lagging behind schedule. This has happened for reasons like the lack of planning powers or fragmented land ownership. Yet on page 81 it is presented without reference to the mixed records, and it appears to non-expert readers as a new policy. Furthermore, on the same page, road widening is presented as an important task for providing escape and evacuation corridors. However, a revealing image (Fig. 10.1) is chosen to represent the policy: In the absence of strong government authorities and owing to a reluctance to exercise imminent domain, it takes decades for the TMG to acquire the necessary land for a road building project, on a plot by plot basis. As the image shows, every new plot that is acquired is fenced in and asphalted over in order

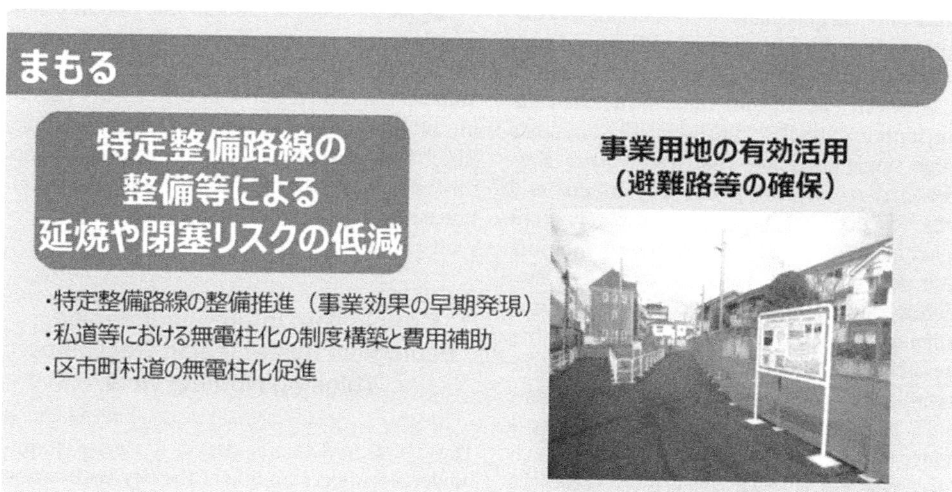


**Fig. 10.1.** One important disaster mitigation measure is the construction of evacuation roads. These have to be cut through the existing city. Assembling land takes many years and parcels purchased by the government are fenced in and asphalted. From Tokyo Resilience Project (TMG (Tokyo Metropolitan Government), 2023a).

to reduce the maintenance costs and the insurance risk for the government and to render these pieces of land inaccessible.

This might make sense from a budgeting, insurance and maintenance perspective, but creating more hard, asphalted surfaces in a city further exacerbates the already pronounced heat island effects and radiative heat. Furthermore, overall, the green coverage ratio is still decreasing in Tokyo and the city is confronted with ever hotter summers, more tropical nights, more frequent flash floods and a sharply rising number of heat-induced health emergencies (Alt, 2023).

It is these inconsistencies between stated goals and realities on the ground that stand out. Another critique is that many featured projects are conceived as single-bottom-line solutions to a supposedly clear and 'tame problem'. However, as Tokyo is confronted with diminishing biodiversity, steadily increasing temperatures, rapid aging and demographic shifts, more frequent and unparalleled torrential rain events and so forth, the above policy might make the city stronger in times of earthquakes but weakens Tokyo as a 'heat-resilient city' or as a 'sponge city'. Thus, many of the presented projects are isolated measures against singular challenges that are hardly coordinated with one another. They promise to create resilience for one isolated aspect while exacerbating vulnerabilities in a number of other domains, without acknowledging this. This suggests a lack of agility in respect to climate change factors.

However, aside from these criticisms, the contribution of the TRP is to list projects together in one single document that were earlier scattered across different policy documents and plans, carried out by disparate government departments and authorities. By combining these projects in one coherent document and placing them in an explicit sustainability and resilience context, one important step has been achieved. Furthermore, by combining the diversified project timelines into one single time frame, provided by the TRP, there might be a greater urgency to implement the different parts of the plan, but whether this is more successful than in the past remains to be seen.

A more cynical view would be that the plan consists of disjointed projects that had long been in the pipeline of different authorities and organisations for their own reasons and objectives, and they are merely listed to demonstrate to the citizens and outside observers the government's commitment to deliver a safe and secure future 'for the next 100 years'.

The plan makes no mention of any concrete coordination mechanisms between the different agencies that are planning and carrying out the discrete respective projects. While the overall plan pays tribute to the challenges posed by climate change, the different projects, such as new disaster evacuation roads or giant underground stormwater drains and storages, display insufficient consideration for long-term sustainability of a society that will lose over 30% of its overall population within the next 50 years, with a share of people older than 65 years near 40% (MHLW, 2025). How will the maintenance costs for these giant infrastructure projects be covered as Japan rapidly ages and depopulates while it is already today confronted with a debt-to-GDP ratio of 263%? How will Tokyo's dynamic development, with selective depopulation and hyper-ageing of some parts of the city and population growth and large renewal projects of other areas, affect the performance of such rigid engineering structuring? Since these are uniquely complex and dynamic challenges, concrete solutions cannot be expected in the short run. However, the Tokyo Metropolitan Government should communicate these issues to the citizens and invite national and international experts, non-profits and other civil society organisations into an ongoing future dialogue. Moreover, a return to the system theory thinking of the 1960s and 1970s (see for example Rittel and Webber, 1973) promises to develop more efficient and emergent networked adaptation strategies.

## Conclusion: Interdependencies, Ecological Systems and Resilience Thinking Nothing New

The above reflections depict a rather static, modernist understanding of the city, with socio-ecological change, and by extension resilient outcomes, being competently managed and controlled by experts and policy makers.

Without elaborating too deeply on the term 'resilience', contrasting some understanding of its use for global cities is useful. Jane Jacobs, over 60 years ago, in her 1961 work *The Death and Life of Great American Cities*, displayed a profoundly ecological understanding of cities, with complex interdependencies between different social, economic and cultural subsystems. In this sense, her book may be seen as representing an urban version of Rachel Carson's bestseller *Silent Spring* (1962), which demonstrated the long causal chains that connected biodiversity, nature, health, human activity and cultural values. Both Jacobs and Carson emphasised a complex, interdependent ecology that constantly changed, emerged and adapted over time: all of which is critically pertinent to understanding cities. This context is very different from the static understanding of a city, as displayed in the TRP.

Furthermore, Jacobs (1961) had identified the nexus between government departments, local politics, governance and planning as critical for determining the success or failure of urban policies. In the chapter 'Governing and planning districts', Jacobs warns against sub-optimal governance, fragmented public policy delivery and segmentation of knowledge and agencies working from within different vertically and horizontally separated silos (1961, pp. 405–427). Interestingly, the Japanese term *tatewari gyōsei*, or 'vertically fragmented administration', describes exactly this issue. Many planning and governance practitioners in Japan have been seeing this fragmentation as one of the root causes for disjointed and sometimes contradictory urban policies (Tamura, 2006). Jacobs also stresses the social and economic as well as spatial resilience and adaptability of an organically and incrementally emerging city to external shocks that stands in contrast to the vulnerability of a modernist and functionalist city that ignores complex interdependencies.

Similarly early, Rittel and Webber (1973), originally 1969 pointed out that 'the kinds of problems that planners (and policy makers) deal with – societal problems – are inherently different from the problems that scientists ... deal with' (1973, p. 160). They stress that planning problems are inherently 'wicked', as 'distinguished from problems in the natural sciences, which are definable and separable and may have solutions that are findable'. Contrary to these are the inherently messy 'problems of governmental planning – especially those of social or policy planning – [that] are ill-defined; and [that] rely on elusive political judgment for resolution' (1973, p. 160).

The difference between 'wicked problems' that we find in the real world, as opposed to 'tame problems' that policy makers and planners seek to tackle, is that there are 'links tying open systems into large and interconnected networks of systems, such that outputs from one become inputs to others ... [In such contexts] it is less apparent where problem centres lie, and ... where and how we should intervene even if we do happen to know what aims we seek [to address]' (1973, p. 159).

The TRP seeks to tackle several viable policy problems through a seemingly impressive range of discrete solutions. However, a more dynamic, ecological understanding of a city is absent, with solutions to some isolated issues creating new vulnerabilities and a reduced resilience in other domains.

The concepts of 'differentiation, resilience, and "bottom-up" ... reflect Jacobs' ecological understanding that systems evolve and adapt according to the specific contexts in which they are situated' (Rowe, 2014, p. 27). Jacobs saw a city's diversity, with redundant subsystems and empowered citizens, able to self-organise in the face of rapid change, all as a prerequisite for constantly adapting to a changing world. 'A system is at its most diverse, and differentiated, at its extremities – where it is most local. That is its strength – the dense weave of interdependencies' (Rowe, 2014, p. 27).

The Tokyo Metropolitan Government initially released the TRP in 2022, followed by an 'upgrade' 1 year later (TMG (Tokyo Metropolitan Government), 2023a). Interestingly, the Japanese title of the policy does not use the katakana version (a script primarily used in Japanese for words of foreign origin) of the English word 'resilience' but the Japanese term *kyōjin* (強靭), which translates to 'strong' and 'tenacious' in everyday language.

In engineering jargon, however, *kyōjin* is used to describe a solid state, combining strength, in the sense of hardness, with toughness. Strength is critical to resisting deformation and fracture in physical systems. However, it is important to remember that actual stability against fracture is not determined by strength alone, but also in resisting propagation of impacts, which is characterised by robustness. The message conveyed by the *kyōjin* image in the TRP is a rather static one. Strength is engineered and tailored for a spectrum of clearly defined scenarios and forces. As the world changes, because of climate change or other stresses, and new, unanticipated scenarios emerge, such an operational design structure is unlikely to cope and evolve.

This position between resistance to change and adaptation to it seems an echo of Holling's (1973) contrasting position between engineered resilience and ecological resilience. Although the concept of 'resilience' has grown into mainstream usage over the past 15 years, it remains somewhat vague. How we assess the resilience of a city or place depends on many tacit assumptions. Where do we draw the line between one city and the next? How do we account for demographic shifts overall and in different parts of the city? How do we build coping and governance capacities, when we cannot fully predict how climate change will locally manifest and be compounded by other crises? In the TRP there is little reference to how Tokyo would coordinate plans and disaster responses with neighbouring cities and prefectures. Also, coordination between different local government entities within the Tokyo Metropolis is hardly mentioned in the TRP.

Clearly, the leading focus here is on engineering solutions, while governance capacities seem of little interest in the TRP document. This too is a recurring theme in urban governance across Japan. Key questions warrant attention. Namely, what kind of extreme events should a city be resilient against? What conceptual approach to resilience is needed or being applied in the first place?

Is it enough if a city 'bounces back' as quickly as possible to resume functioning in the manner it did pre-disturbance? Does consistency exist in the application of resilience that emphasises an adaptive, emergent capacity?

One could argue that the initial vision of Gotō Shimpei would have created a much greener, more resilient city, far better suited to cope with anticipated future challenges that would emerge over 100 years later. Yet these concerns were not broadly shared by the ruling elites and the citizens, and thus much of his visionary planning was scrapped, arguably due to funding priorities and power retention among different political centres, as well as competing jurisdictions between ministries and government authorities. In other words, the intent was application of engineered robustness and reliability that sought a return of a system to specific, narrowly defined functions and performance criteria, not the notion of ecological resilience that envisions the emergence of different states of equilibrium (Holling, 1973), better suited to a changing socio-ecological world context. Gunderson points out that appreciating 'multiple meanings of resilience can lead to very different options for policies and actions' (Gunderson, 2000, p. 425).

The TRP operates mostly as a technical, expert-centred, top-down, engineer-focused project. Engineers are seen in a promotional video in front of large screens and control panels in a situation room, operating complex disaster prevention infrastructures.[3] The following sequence of the video stresses that new technological innovations like AI will further improve the expert-driven disaster management system. The subtext is that disasters *can* be accurately modelled and simulated, and appropriate plans and projects *can* be devised to minimise damage to property and loss of life. To a degree, this is entirely possible, but a more open-ended, emergent view of the world could create stronger, more deeply rooted forms of preparedness.

The TRP documentation contains little to no reference to active citizen participation in democratic governance. A noted asymmetry is at play between experts and non-expert citizens. It is noteworthy that in the TRP promotional video, citizens only appear as passive props who enjoy a carefree, private life in streets and green parks, while experts in faraway control rooms are the invisible guardians that guarantee safety. Interestingly, in safety videos in Korea, often citizens themselves operate emergency infrastructure, provided by the government, and help each other out of difficult situations.[4] The difference couldn't be more striking. Admittedly, an expert-driven approach has its advantages, as decisions can be more efficient and swifter in response to an imminent disaster. However, as I wrote in a *Japan Times* column in September 2011, shortly after the Great East Japan Earthquake, tsunami and nuclear disaster,

> a key issue for the creation of resilient cities is the presence of a healthy civil society and a vital public sphere, where alternative futures are broadly debated – not just by a few experts. Rebuilding should therefore be seen less as an end in itself, but instead as a continuous process through which civil society develops more fully, communities can grow again closer, and the entire country can become more resilient and self-reliant. Any reconstruction proposal must be measured in terms of its tangible outcomes, of course, but also by the extent to which it contributes to the empowerment, political participation and social inclusion of the citizenry. (Dimmer, 2011)

A rapidly growing body of research is confirming that in post-disaster scenarios it is social capital (Aldrich, 2012, 2019) and community resilience (Aldrich and Meyer, 2015; Dimmer and Kammerbauer, 2024) that play a determining role in the success of recovery and reconstruction and for future disaster preparedness. 'Soft' social capital and community resilience are at least as important as hard, urban infrastructure.

Finally, how we evaluate resilience and how urban resilience is achieved depends to a large degree on competing visions of society, who holds forms of power in that society and how the government positions itself vis-à-vis the citizens. As our future natural hazard landscape evolves, city resilience will depend on how the governed and those who govern see the world. Do we see the world as stable and manageable or as constantly changing and emergent and evolving forms? Do we believe in the infallibility of experts, or do we invite citizens into the making of preparedness and resilience strategies? Do we believe in a democratic, pluralistic and participatory society in which citizens and their representatives cooperatively search for workable adaptation solutions? Or should our society be one in which a passive citizenry engages mostly in private matters and relies on expert engineers to steer socio-technical change? Abstract as these questions may appear, they play a crucial role in evaluating how appropriate a resilience strategy for a given city really is.

## Notes

[1] In Japanese, the term 'disaster prevention' is more frequently used than 'disaster mitigation'. Thus, underlying is a notion that the impacts of disasters can be prevented. This suggests a very strong belief in the capabilities of engineering and technology.

[2] https://tokyo-resilience.metro.tokyo.lg.jp/kanto-daishinsai/ (accessed 1 May, 2024).

[3] Tokyo Metropolitan Government (2023). Promotional video Tokyo Resilience Project (Tokyo Kyoshinka Purojekuto), minute 2:16–2:50: https://www.youtube.com/watch?v=aBWIl2Ufy80 (accessed 1 May, 2024).

[4] See for example a fire evacuation video of the Seoul Metro, in which no officials appear and active and trained citizens take care of their own evacuation:
http://www.seoulmetro.co.kr/kr/board.do?menuIdx=555&bbsIdx=2214673 (accessed 1 May, 2024).

## References

Aldrich, D.P. (2012) *Building Resilience – Social Capital in Post-Disaster Recovery*. University of Chicago Press, Chicago, Illinois.

Aldrich, D.P. (2019) *Black Wave – How Networks and Governance Shaped Japan's 3/11 Disasters*. University of Chicago Press, Chicago, Illinois.

Alt, M. (2023) The shock of Japan's extreme heat. The New Yorker. Available at: https://www.newyorker.com/culture/cultural-comment/the-shock-of-japans-extreme-heat (accessed 16 July 2025).

Brown, P.C. (2014) Call it a 'wash'? Historical perspectives on conundrums of technological modernisation, flood amelioration and disasters in modern Japan. *The Asia-Pacific Journal* 12(7), 2.

Aldrich, D.P. and Meyer, M.A. (2015) Social capital and community resilience. *American Behavioural Scientist* 59(2), 254–269.

Carson, R. (1962) *Silent Spring*. Houghton Mifflin Harcourt, Boston, Massachusetts.

Clancey, G. (2006) *Earthquake Nation: The Cultural Politics of Japanese Seismicity, 1868–1939*. University of California Press, Oakland, California.

Dimmer, C. (2011) Restructuring for the future, not rebuilding the past. *The Japan Times,* Hotline to Nagatacho, 20 September.

Dimmer, C. (2018) Japan after March 11th 2011: Between swift reconstruction and sustainable restructuring. In: Yan, W. and Galloway, W. (eds) *Rethinking Resilience: Adaptation and Transformation in a Time of Change*. Springer, Cham, Switzerland, pp. 23–40.

Dimmer, C. and Kammerbauer, M. (2024) Sustainability and adaptation in planning: Community resilience against accelerating environmental change. In: Hommerich, C. and Kimura, M. (eds) *Sustainability in a Fragile World – Approaches from Germany and Japan*. Sophia University Press, Tokyo, pp. 95–122.

Fujisaki, T. (2023) Tokyo resilience project. In: *Presentation at the UNECE Urban Resilience Week*, Geneva, Switzerland, 2–6 October 2023. Available at: https://unece.org/sites/default/files/2023-10/2_Tetsuro_Fujisak.pdf (accessed 1 May 2024).

Gunderson, L. (2000) Ecological resilience – in theory and application. *Annual Review of Ecology and Systematics* 31, 425–439.

Hein, C. (2005) Resilient Tokyo – disaster and transformation in the Japanese city. In: Vale, L.J. and Campanella, T.J. (eds) *The Resilient City - How Modern Cities Recover from Disasters*, Campanella. Oxford University Press, Oxford, UK, pp. 213–234.

Holling, C.S. (1973) Resilience and stability of ecological systems. *Annual Review of Ecology, Evolution, and Systematics* 4, 1–23.

Jacobs, J. (1961) *Death and Life of Great American Cities*. Random House, New York.

Kelly, W.W. (1994) Incendiary actions: Fires and firefighting in the Shogun's capital and the people's city. In: McClain, J.L., Merriman, J.M. and Ugawa, K. (eds) *Edo and Paris: Urban Life and the State in the Early Modern Era*. Cornell University Press, Ithaca, New York, pp. 310–331.

Krolicki, K., DiSavino, S. and Fuse, T. (2011) Japan engineers knew tsunami could overwhelm Fukushima plant. Reuters, 30 March. Available at: https://www.reuters.com/article/business/energy/special-report-japan-engineers-knew-tsunami-could-overwhelm-fukushima-plant-idUSL3E7EU0HO/ (accessed 16 July 2025).

McClain, J. (1994) Edobashi: Power, space and popular culture in Edo. In: McClain, J.L., Merriman, J.M. and Ugawa, K. (eds) *Edo and Paris: Urban Life and the State in the Early Modern Era*. Cornell University Press, Ithaca, New York, pp. 105–131.

MHLW (Ministry of Health, Labour and Welfare) (2025) 2025 White paper on aging society (Full Version). Available at: https://www8.cao.go.jp/kourei/whitepaper/w-2025/zenbun/07pdf_index.html (accessed 1 August 2025).

Nippon.com (2019) The changing face of Tokyo: One in eight Shinjuku residents are foreign nationals. Available at: https://www.nippon.com/en/japan-data/h00398/the-changing-face-of-tokyo-one-in-eight-shinjuku-residents-are-foreign-nationals.html (accessed 1 May 2024).

Onishi, N. (2011) Japan revives a sea barrier that failed to hold. New York Times. Available at: https://www.nytimes.com/2011/11/03/world/asia/japan-revives-a-sea-barrier-that-failed-to-hold.html (accessed 16 July 2025).

Rittel, H. and Webber, M. (1973) Dilemmas in a general theory of planning. *Policy Sciences* 4(2), 155–169. Originally presented to Panel on Policy Sciences, American Association for the Advancement of Science, Boston, December 1969.

Rowe, M. (2014) Jane jacobs and the self-organising city. In: Schubert, D. (ed.) *Contemporary Perspectives on Jane Jacobs*. Ashgate, Farnham, UK, pp. 21–30.

Schencking, C. (2013) *The Great Kantō Earthquake and the Chimera of National Reconstruction in Japan*. Columbia University Press, New York.

Scott, J.C. (1998) *Seeing Like a State: How Certain Schemes to Improve the Human Condition Have Failed*. Yale University Press, New Haven, Connecticut.

Seidensticker, E. (1991) *Tokyo Rising – The City Since the Great Earthquake*. Harvard University Press, Cambridge, Massachusetts.

Sorensen, A. (2002) *The Making of Urban Japan: Cities and Planning from Edo to the Twenty First Century*. Routledge, Abingdon, UK.

Tamura, A. (2006) *The Struggle of City Planner Tamura Akira: Towards a Citizen's Government in Yokohama (Toshi Puranna Tamura Akira no Tatakai: Yokohama Shimin no seifu wo mezashite)*. Gakugei Shuppansha, Kyoto, Japan.

Tate, J. (2009) Fights and fires are the flowers of Edo. *The Design Journal* 12(1), 9–24.

TMG (Tokyo Metropolitan Government) (2023a) Tokyo resilience project – upgrade 1. Available at: https://www.spt.metro.tokyo.lg.jp/seisakukikaku/tokyo-kyoujinka/trp_upgrade/ (accessed 1 May 2024).

TMG (Tokyo Metropolitan Government) (2023b) Tokyo sustainability action 2023. Available at: https://www.english.metro.tokyo.lg.jp/documents/d/seisakukikaku/tokyo-sustainability-action-2023 (accessed 1 May 2024).

TMG (Tokyo Metropolitan Government) (2023c) Tokyo metropolitan public relations bulletin, February 2023 Issue: Launch of 'SusHI Tech Tokyo' (Sustainable High City Tech Tokyo). Available at: https://www.koho.metro.tokyo.lg.jp/2023/02/01.html (accessed 1 May 2024).

Weisenfeld, G.S. (2012) *Imaging Disaster: Tokyo and the Visual Culture of Japan's Great Earthquake of 1923*. The University of California Press, Oakland, California.

# Conclusion: Cloudy with 'More' than a Chance of Chaos

**Paul Barnes***
*HEAL Senior Research, Policy and Practice Fellow, Centre for Environment and Population Health, School of Medicine and Dentistry, Griffith University, Australia*

---

The phrase 'Cloudy with "more" than a chance of chaos' plays on the familiar weather forecast trope 'Cloudy with a chance of rain.' It is used here metaphorically to describe a complex landscape of contemporary global challenges where climate change, technological disruption and social transformation have rendered conventional approaches to problem-solving less effective at best and in the extreme, obsolete. It also establishes a view that uncertainty is rampant, and disruptions are not a matter of possibility, but increasingly likely or even inevitable.

Critical incidents (or multiple concurrent incidents: polycrises) may emerge unexpectedly at any time in large, highly complex systems, or they may 'cook' slowly (with limited recognition) until some triggering event or process precipitates them. In either case, disruptions can be both surprising and unexpected. Surprise has always had an egalitarian effect in society. To be surprised in a pleasant fashion is preferable to the alternative. Unfortunately, the alternative state seems more likely through time.

Emergent complexities in linked systems often make the impact of emergencies difficult to anticipate and consequences difficult to plan for. Further, under emergency conditions, the pressure on decision takers to 'make-sense' of multiple lines of information (in both crisis and consequence contexts) is significant. Equally important is the question of how lay members of society make sense of their vulnerability to the harm and impacts that may be likely in a chaotic future.

The 10 chapters of this compilation explore a selection of viewpoints on complex challenges humanity faces in dealing with climate change and natural hazards. Key themes stand out:

- that the complexity of climate-driven change demands new approaches that match this complexity – not just in technical sophistication driven by extended or refined specialities alone, but in an intentionally nurtured ability to acquire and integrate diverse types of knowledge and expertise while maintaining flexibility and adaptiveness in response capabilities and recovery systems;
- that successful climate adaptation requires transforming how we think about and implement resilience: moving beyond engineering-based approaches to embrace more holistic frameworks that recognise the dependencies and interdependencies of social, technical and environmental systems; and

*Corresponding author: paul.barnes@griffith.edu.au

 *Climate Change and Risk Mitigation: Reducing Vulnerabilities and Enhancing Resilience* (ed. P. Barnes)
DOI: 10.1079/9781800624153.conc

- that (arguably) pathways forward require balancing seemingly contradictory elements such as expert knowledge with local wisdom, technological innovation with social adaptation and centralised coordination with community autonomy. Options for effective decision making are not based on choosing between these alternatives but in finding ways to integrate them effectively.

In Chapter 1, Barnes positions climate-driven disruptions in the broad context of cascading 'polycrises'. He argues that addressing these complex problems requires breaking down traditional siloed thinking in favour of multi- and transdisciplinary approaches that map the complexity of the challenges to a matched complexity and depth of capability that enhances options for response. This draws on Ashby's Law of Requisite Variety, which has been generically expressed as 'only variety can absorb variety'.

Barnes proposes a framework for crisis management that maps the progression of crises through generic stages: preconditions and triggering events, emergent issues or disturbances, primary disruptions/impacts and secondary/tertiary impacts leading to ongoing disruptions. For each stage, he outlines possible actions and the types of expertise needed, emphasising that successful crisis management requires integration of diverse perspectives from systems analysts, social scientists, natural scientists, policy experts, technologists and communication specialists.

The chapter stresses that multidisciplinary teams must employ integration mechanisms including shared conceptual frameworks, structured decision-making processes and clear operational protocols that allow different disciplines to contribute while maintaining coherence in approach and efficiency in dealing with crisis situations and subsequent disruption. Of course, combinations of technical expertise applied in effective ways are not enough. The *what* and *how* of skill combination is important, but understanding and defining *why* is equally critical.

In Chapter 2 De Marchi and Menoni examine why diverse knowledge integration for complex environmental and disaster challenges delivers value by defining the importance of stepping past traditional scientific approaches to handle problems that are characterised by uncertain facts, disputed values, high stakes and urgent decisions: in this case, flood damage assessment and mitigation planning in the Umbria region of Italy.

Basing analysis within a post-normal science frame, where researchers and local authorities collaborate to develop better systems for post-disaster assessment, this 'living lab' experience demonstrates how different forms of knowledge – scientific, legislative, organisational and local – can be successfully integrated to create more effective tools and procedures.

De Marchi and Menoni show that the damage assessment effort, and logically, disaster risk reduction processes, benefitted from using a wider focus than just technical assessment, by including value judgements and multiple stakeholder perspectives, all of which lead to effective and comprehensive results. The broader implications of these findings extend across multiple domains, from adapting to climate variability and risk reduction to public health emergencies. It is argued that institutions must develop adaptive and agile governance structures capable of functioning in real-world settings where irreducible uncertainty and the criticality of making timely and useful decisions are to the fore.

As suggested in Chapter 1, breaking down traditional boundaries between disciplines and sectors is a critical multiplier of effective governance in complex settings. A core message is that policy makers and technical experts should invest in creating and empowering 'extended peer communities' before crises occur and 'bake in' the inclusion of local knowledge (and by implication, their interests) as critically legitimate and useful inputs to decision making.

In Chapter 3, Crosweller extends the institutional/community nexus in a landscape of escalating climate disasters to suggest that leadership – in pre-, during and post-disruption settings – requires rethinking the social contract between governments and citizens based on a relational ethical approach balancing principles and contextual understanding.

He suggests disaster management leaders need to develop 'relationality' by becoming open to emotional engagement, be willing to change systems, become tolerant of differences and be able to genuinely empathise with (community) suffering and loss. This relational approach, he argues, should navigate between principle-based ethics and ethical frames that recognise local context and values.

Crosweller introduces three factors to help leaders better understand community needs: sentiment (understanding emotional responses to loss and exposure to harm), symbolism (finding representative stories that connect leaders to local experiences and suffering) and data (validating these understandings analytically). This balanced approach aims to strengthen trust while reducing suffering within a social contract of shared responsibility. In severe to catastrophic disasters, both governments and citizens become mutually vulnerable in differing ways. Two key ethical questions arise: how to act in ways that uphold trust and confidence, and how to best negate, minimise or manage avoidable suffering, loss and deepened vulnerability to future disruptions.

Crosweller concludes by emphasising the need for disaster management leaders to adopt a relational approach, informed by mutual trust and compassion. This then provides a stability to effectively navigate the challenges posed by the social contract of shared responsibility in the context of an increased frequency of climate change-driven disasters and services disruption.

In Chapter 4, Fien and Charlesworth examine climate-related disasters from a systems thinking perspective. They position disasters as complexly adaptive 'wicked problems' requiring holistic systemic approaches that are not limited by conventional methods. Their classification of disasters as 'wicked problems' is apt in that disruptive effects, recovery and remediation options often have no clear definition or solution and are constantly changing, and attempted solutions may create new problems.

In their view, anticipating the (networked) impacts of disasters has become increasingly complex in the Anthropocene era, with the landscape characterised by interconnected 'polycrises' where multiple challenges and disruptions 'compound and reshape' each other. They emphasise the importance of approaching 'interventions' for response and recovery as iterative examinations (experiments) that need to be monitored for unintended consequences.

The authors suggest that addressing these complex disaster exposures requires moving beyond engineering resilience (often focused on returning systems to their original state) towards socio-ecological and transformative approaches that can fundamentally reshape systems to be more sustainable and equitable.

They advocate for approaches to disaster risk reduction that combine systems thinking (deeply cognisant of complex adaptive systems) with design thinking, following seven key steps: framing the system, listening to the system, understanding the system, defining desired futures, exploring possibilities, designing intervention strategies and implementing/monitoring transition plans. Such an approach emphasises collaboration, creative thinking and treating all interventions as experiments to be carefully monitored and adjusted.

Sustainable energy transitions at a regional and city level are central elements of climate change adaptation. In Chapter 5, Roux and Shearing examine the transition of Cape Town's electricity regime to renewable energy sources through the lens of value systems and social contracts, challenging traditional socio-technical transition theories. Cape Town has committed to achieving carbon neutrality by 2050, embedding this goal in its Integrated Development Plan and aligning its process with international climate agreements.

Using historical analysis, they identify four distinct periods of regime change in the development of Cape Town's electricity system, each driven by shifts in underlying societal values: the initial experimental period (1879–1912), the apartheid era of centralised control (1912–1994), the post-apartheid focus on development and justice (1994 onwards) and the current transition towards environmental sustainability (1998 onwards). The authors detail how these transitions were shaped by complex interactions between policy networks, social movements and changing value systems rather than purely technological or market forces.

A key finding was the existence of a tension between developmental values (focused on equity and access) and environmental values (focused on sustainability) in Cape Town's electricity regime. Roux and Shearing detail how the city's electricity revenue model, which used profits from electricity sales to cross-subsidise services for poor communities, created institutional resistance to environmental initiatives like renewable energy and energy efficiencies that could reduce electricity consumption.

Chapter 6, by Okokon, Tong and Vardoulakis, examines the environmental health impacts of increasing heat and humidity likely to accompany climate change. They detail how rising global temperatures increase atmospheric moisture capacity, creating more extreme precipitation events while simultaneously amplifying heat exposure through higher humidity that impedes the body's natural cooling mechanisms. The authors explain how these conditions stress human thermoregulation, particularly when wet-bulb temperatures approach 35°C, beyond which the body cannot effectively cool itself.

Urban areas face compounded risk exposure due to 'heat island' effects, with city temperatures that become significantly higher than surrounding urban-rural areas due to reduced vegetation, heat-absorbing building materials and waste heat from building systems. These conditions lead to immediate health impacts (heatstroke, exhaustion, cramps) and exacerbate chronic conditions, while increasing cooling energy demands that further strain resources and deepen energy inequities.

The chapter outlines adaptation strategies that require multi-sectoral collaboration, including the redesign of infrastructure, enhanced health system preparedness and social support networks. Key approaches suggested include the expansion of urban green spaces for cooling through evapotranspiration; implementing passive cooling building designs that reduce energy consumption; developing early warning systems for extreme heat events; establishing cooling centres for vulnerable populations; and strengthening healthcare provider training for heat-related illnesses.

In Chapter 7, De Leeuw and Green analyse how multiple global crises, including climate change, require cities to develop effective governance strategies supportive of sustained resilience. They highlight the growth of various 'City' themed networks (including Healthy Cities, Resilient Cities and Age-Friendly Cities) as responses to these challenges, and criticise how even recent health-focused climate resilience frameworks can fail to properly integrate urban dimensions.

De Leeuw and Green further explore the idea that resilience should be understood as more than just 'bouncing back' from adverse events. Drawing on ecological principles and extant literature on resilience, the authors emphasise a need to differentiate between three extant components of resilience-related mechanisms: disturbance, response and outcome. They note that promoting resilience in urban environments assumes an understanding and appreciation of the inequities of the impacts of climate-related events on different urban populations, particularly in relation to social infrastructure and community capacity.

The authors point to the Age-Friendly Environments in Europe framework as a proven, integrated approach that could be applied to climate resilience, demonstrating how investment in both viable physical environments and social infrastructure can enable cities, as human-centric systems, to respond effectively to challenges like the COVID-19 pandemic.

A range of climate-driven impacts have been associated with food supply security and agricultural productivity in peri-urban settings, particularly in Africa. In Chapter 8 Ackermann and Faragher present an example of community-level adaptation to these pressures in the form of the Langa Agrihub gardens in Cape Town, South Africa.

Ackermann and Faragher detail how the Agrihub model supports a network of urban farmers in historically disadvantaged communities, where residents face high unemployment, poverty and food insecurity, by providing affordable inputs, training, market access and social infrastructure. While the authors acknowledge that urban food gardens cannot fully resolve food insecurity, they document multiple positive outcomes, including economic improvements (increases in harvest value), enhanced social cohesion, health benefits, climate adaptation capabilities and civic engagement.

The chapter affirms that Agrihubs can function as effective social infrastructure that strengthens community resilience against various shocks, including climate disasters, transportation strikes and emergent disease.

In Chapter 9, Kesavan, Samhita and Sharma detail a Delhi case study of engagement with communities in the Vivekanand Camp using artificial intelligence tools that enable targeted interventions for heat-vulnerable groups among camp communities by providing detailed local-level heat risk assessments.

Unlike traditional heat warning systems that offer blanket advisories for entire regions, the ResSolv AI-based tool generates risk assessments at a building level. An integration of satellite imagery, analysing building materials, roof typologies and spatial layouts, allows the identification of specific hotspots across urban areas. The authors highlight how urban heat impacts are not experienced uniformly across cities: particularly where neighbourhoods exhibit low-income, dwellings are densely built and lack green spaces, have building materials that trap heat and have limited ventilation.

The granular risk assessments created by the ResSolv system enable targeted retrofitting using locally available and recycled materials that have been shown to reduce indoor temperatures by up to 12°C. The system also enables customised warning systems for vulnerable communities and families, enhancing agency and choices at scales from the family level upwards.

In Chapter 10, Dimmer moves the focus to adaptation to climate change pressures in modern megacities. He examines Tokyo's shift from historical community-based 'disaster culture' to one that is predominantly expert-driven with an engineering-focused model of risk reduction, contrasting pre-modern and modern perspectives through the lens of the recently launched Tokyo Resilience Project (TRP).

In pre-modern Edo/Tokyo, disasters were accepted as inevitable, leading to the development of a 'disaster culture' that emphasised swift recovery and community-level resilience. However, with modernisation, this approach has shifted towards an expert-driven, engineering-based approach that prioritises technological solutions and planned infrastructure projects and potentially reduces overall societal resilience by disengaging citizens from active participation in disaster preparedness and response activities.

An analysis of the TRP reveals several critical limitations in Tokyo's current approach to resilience building. While the project includes an impressive array of infrastructure initiatives, it lacks meaningful citizen engagement, fails to address governance coordination challenges and doesn't adequately consider the compound effects of climate change alongside demographic challenges like rapid aging and depopulation.

Dimmer argues that the TRP's engineering-focused approach to resilience (*kyōjin* or strength/tenacity) may be too rigid to provide a socio-technical adaptability needed to address the complexity of the disturbances that can expected within a megacity from climate-driven crises.

## What Then Might 'We' Need To Do?

The chapters in this collection examine how institutions and societies might better prepare for and respond to the complex, interconnected challenges posed by climate variability and natural hazards. A central theme emerging across contributions is that traditional siloed approaches to risk assessment and management have become increasingly inadequate for addressing the cascading and compounding nature of contemporary climate-driven disruptions.

A number of key insights emerge from the collected works. There is growing recognition that threats from climate variability cannot be understood or addressed in isolation – they are inherently systemic and interconnected, requiring new frameworks that can capture complex dependencies, interdependencies and feedback loops across social, technical and environmental domains. Central to these points is how to how to make sense of a lived experience within such complexity.

There is an urgent need to develop more complete and inclusive approaches for applying transdisciplinary capabilities, particularly in time- and resource-constrained settings. These include practical frameworks for integrating knowledge, metrics for evaluating the effectiveness of collaborative approaches and guidelines for incorporating diverse forms of expertise.

Evolving technologies applying artificial intelligence and machine learning techniques are likely to offer new opportunities to assess and enhance understanding of the complex dynamics of disturbed socio-technical-environmental systems and for anticipating potential cascading impacts. However, these must be developed thoughtfully, with careful attention to issues of equity and inclusion.

Finally, there is growing recognition that effective climate adaptation requires fundamental transformation and rethinking of how institutions and governance systems can better match the complexity and interconnectedness of contemporary challenges, rather than just get better at incremental change. A sustained commitment is needed to invest in breaking down siloed policy and practice, building 'bridges' across disciplines and sectors, and for developing new capabilities for collaborative issues definition and problem-solving across multiple scales.

This means not just bringing different experts together, but developing new ways of synthesising diverse forms of knowledge – from scientific expertise to local and first nations knowledge – to create a more complete understanding of current and future challenges as well as solutions.

Thinking about the future can be imbued with metaphor. The notion of 'expecting the unexpected' is convenient but also facile, with limited usefulness. A more useful positioning, arguably, is to anticipate disruptions that can emerge in known and knowable forms: this is where the polycrisis construct seems useful. While still evolving as a concept, it offers a valuable lens for understanding how multiple crises emerge and disrupt, interact with and amplify each other's effects.

A viable strategy may begin with incentivising collaboration on agile policy development and decision making (across local and regional settings and in national and international institutions) and enhancing awareness of the complex disturbances that can occur. This forms a future-focused type of continuity thinking that can support effective recovery of essential societal functions though time. A structural representation of the core themes of such a strategy is depicted in Fig. C.1.

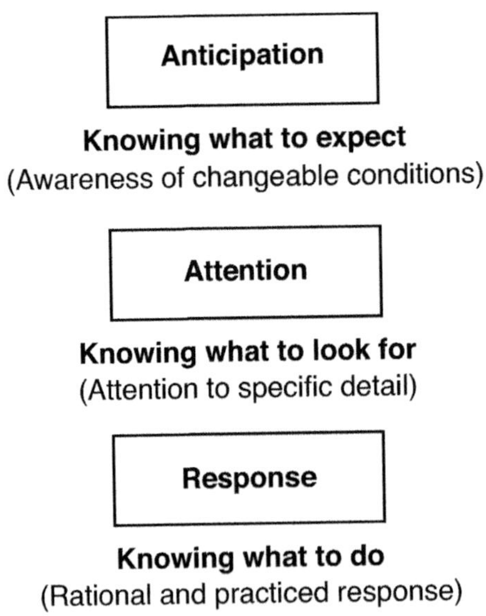


**Fig. C.1.** Anticipation strategy. Derived from Hollnagel and Woods, 2005, after Diamond, 2005.

Knowing what disturbances can occur and being able to detect the triggers for their emergence is critical, as is knowing how to respond to and mitigate their effects and initiate recovery. While such capabilities require a whole-of-society effort, it is at the local and regional level that effective continuity management must be implemented.

However, both preventing and preparing for multiple crisis situations presumes a deep and effective understanding of the way 'unexpected' factors and conditions can manifest and how they might directly or indirectly exploit organisational and institutional vulnerabilities.

Responding to and recovering from crises also assumes an effective appreciation of mitigation and consequence factors. This level of understanding presumes planners, advisors and responders can make sense of confusing and at times conflicting information flows: a presumption that many iconic disasters have shown to be absent.

As we face this era of ubiquitous challenge, perhaps our greatest resource is not technological prowess or institutional resources alone, but a capability and capacity for collective imagination, empathy and collaborative transformation of how we interpret existential threats, their meaning and significance, and how we prepare for and respond to them. Some might say that the future belongs to those who can navigate uncertainty with creativity, collective skills, compassion and a deep commitment to the systemic wellbeing of our selves and the wider lived and biological environments.

## References

Diamond, J. (2005) *Collapse: How Societies Choose to Fail or Succeed*. Viking Books, London.

Hollnagel, E. and Woods, D.A. (2005) Epilogue: resilience engineering precepts. In: Hollnagel, E., Woods, D.A. and Leveson, N. (eds) *Resilience Engineering: Concepts and Precepts*. Ashgate, Farnham, UK.

# Index

Printed and bound by CPI Group (UK) Ltd, Croydon, CR0 4YY
16/09/2026
14963198-0002